Math Challenge I-B
Number Theory

Areteem Institute

Math Challenge I-B Number Theory

Series: Math Challenge Curriculum Textbooks, Vol. 20

Edited by   David Reynoso
            John Lensmire
            Kevin Wang
            Kelly Ren

Copyright © 2018 ARETEEM INSTITUTE

WWW.ARETEEM.ORG

PUBLISHED BY ARETEEM PRESS
ALL RIGHTS RESERVED. No part of this publication may be reproduced, stored in a retrieval system, or transmitted, in any form or by any means, electronic, mechanical, photocopying, recording, or otherwise, without prior written permission of the publisher, except for "fair use" or other noncommercial uses as defined in Sections 107 and 108 of the U.S. Copyright Act.

ISBN: 1-944863-39-7
ISBN-13: 978-1-944863-39-5
First printing, September 2018.

TITLES PUBLISHED BY ARETEEM PRESS

Cracking the High School Math Competitions (and Solutions Manual) - Covering AMC 10 & 12, ARML, and ZIML

Mathematical Wisdom in Everyday Life (and Solutions Manual) - From Common Core to Math Competitions

Geometry Problem Solving for Middle School (and Solutions Manual) - From Common Core to Math Competitions

Fun Math Problem Solving For Elementary School (and Solutions Manual)

ZIML Math Competition Book Division E 2016-2017

ZIML Math Competition Book Division M 2016-2017

ZIML Math Competition Book Division H 2016-2017

ZIML Math Competition Book Jr Varsity 2016-2017

ZIML Math Competition Book Varsity Division 2016-2017

MATH CHALLENGE CURRICULUM TEXTBOOK SERIES

Math Challenge I-A Pre-Algebra and Word Problems
Math Challenge I-B Pre-Algebra and Word Problems
Math Challenge I-C Algebra
Math Challenge II-A Algebra
Math Challenge II-B Algebra
Math Challenge III Algebra
Math Challenge I-A Geometry
Math Challenge I-B Geometry
Math Challenge I-C Topics in Algebra
Math Challenge II-A Geometry
Math Challenge II-B Geometry
Math Challenge III Geometry
Math Challenge I-B Counting and Probability
Math Challenge I-B Number Theory
Math Challenge II-A Number Theory

COMING SOON FROM ARETEEM PRESS

Fun Math Problem Solving For Elementary School Vol. 2 (and Solutions Manual)
Counting & Probability for Middle School (and Solutions Manual) - From Common Core to Math Competitions
Number Theory Problem Solving for Middle School (and Solutions Manual) - From Common Core to Math Competitions
Other volumes in the **Math Challenge Curriculum Textbook Series**

The books are available in paperback and eBook formats (including Kindle and other formats). To order the books, visit https://areteem.org/bookstore.

Contents

	Introduction	9
1	Number Sense and Place Values	13
1.1	Example Questions	14
1.2	Quick Response Questions	17
1.3	Practice Questions	19
2	Primes and Factors	23
2.1	Example Questions	24
2.2	Quick Response Questions	29
2.3	Practice Questions	31
3	Primes and Factors Continued	33
3.1	Example Questions	34
3.2	Quick Response Questions	37
3.3	Practice Questions	39
4	Divisibility	41
4.1	Example Questions	42
4.2	Quick Response Questions	45

Copyright © ARETEEM INSTITUTE. All rights reserved.

4.3	Practice Questions	47
5	**Divisibility Continued**	**49**
5.1	Example Questions	49
5.2	Quick Response Questions	52
5.3	Practice Questions	54
6	**GCD's and LCM's**	**57**
6.1	Example Questions	58
6.2	Quick Response Questions	61
6.3	Practice Questions	63
7	**GCD's and LCM's Continued**	**65**
7.1	Example Questions	65
7.2	Quick Response Questions	69
7.3	Practice Questions	71
8	**Remainders**	**73**
8.1	Example Questions	74
8.2	Quick Response Questions	77
8.3	Practice Questions	79
9	**Modular Arithmetic**	**81**
9.1	Example Questions	82
9.2	Quick Response Questions	85
9.3	Practice Questions	87
Solutions to the Example Questions		**89**
1	Solutions to Chapter 1 Examples	90
2	Solutions to Chapter 2 Examples	99
3	Solutions to Chapter 3 Examples	106
4	Solutions to Chapter 4 Examples	111
5	Solutions to Chapter 5 Examples	116
6	Solutions to Chapter 6 Examples	121

7	Solutions to Chapter 7 Examples	126
8	Solutions to Chapter 8 Examples	132
9	Solutions to Chapter 9 Examples	138

Introduction

The math challenge curriculum textbook series is designed to help students learn the fundamental mathematical concepts and practice their in-depth problem solving skills with selected exercise problems. Ideally, these textbooks are used together with Areteem Institute's corresponding courses, either taken as live classes or as self-paced classes. According to the experience levels of the students in mathematics, the following courses are offered:

- Fun Math Problem Solving for Elementary School (grades 3-5)
- Algebra Readiness (grade 5; preparing for middle school)
- Math Challenge I-A Series (grades 6-8; intro to problem solving)
- Math Challenge I-B Series (grades 6-8; intro to math contests e.g. AMC 8, ZIML Div M)
- Math Challenge I-C Series (grades 6-8; topics bridging middle and high schools)
- Math Challenge II-A Series (grades 9+ or younger students preparing for AMC 10)
- Math Challenge II-B Series (grades 9+ or younger students preparing for AMC 12)
- Math Challenge III Series (preparing for AIME, ZIML Varsity, or equivalent contests)
- Math Challenge IV Series (Math Olympiad level problem solving)

These courses are designed and developed by educational experts and industry professionals to bring real world applications into the STEM education. These programs are ideal for students who wish to win in Math Competitions (AMC, AIME, USAMO, IMO,

---

Copyright © ARETEEM INSTITUTE. All rights reserved.

ARML, MathCounts, Math League, Math Olympiad, ZIML, etc.), Science Fairs (County Science Fairs, State Science Fairs, national programs like Intel Science and Engineering Fair, etc.) and Science Olympiad, or purely want to enrich their academic lives by taking more challenges and developing outstanding analytical, logical thinking and creative problem solving skills.

In Math Challenge I-B, students expand middle school math skills to a deeper level with topics in beginning algebra, fundamental geometry, counting strategies, and basic number theory. The students not only learn practical skills of challenging problem solving that are supplemental to their school curricula, but also develop skills in creative thinking, logical reasoning, oral and written presentation, and team work. This course helps 6th to 8th graders to participate in the American Mathematics Competition (AMC) 8, MathCounts, Math Olympiads for Elementary and Middle School (MOEMS), and Zoom International Math League (ZIML) Division M.

The course is divided into four terms:

- Summer, covering Pre-Algebra and Word Problems
- Fall, covering Geometry
- Winter, covering Counting and Probability
- Spring, covering Number Theory

The book contains course materials for Math Challenge I-B: Number Theory.

We recommend that students take all four terms. Each of the individual terms is self-contained and does not depend on other terms, so they do not need to be taken in order, and students can take single terms if they want to focus on specific topics.

Students can sign up for the live or self-paced course at `classes.areteem.org`.

Copyright © ARETEEM INSTITUTE. All rights reserved.

About Areteem Institute

Areteem Institute is an educational institution that develops and provides in-depth and advanced math and science programs for K-12 (Elementary School, Middle School, and High School) students and teachers. Areteem programs are accredited supplementary programs by the Western Association of Schools and Colleges (WASC). Students may attend the Areteem Institute in one or more of the following options:

- Live and real-time face-to-face online classes with audio, video, interactive online whiteboard, and text chatting capabilities;
- Self-paced classes by watching the recordings of the live classes;
- Short video courses for trending math, science, technology, engineering, English, and social studies topics;
- Summer Intensive Camps held on prestigious university campuses and Winter Boot Camps;
- Practice with selected free daily problems and monthly ZIML competitions at ziml.areteem.org.

Areteem courses are designed and developed by educational experts and industry professionals to bring real world applications into STEM education. The programs are ideal for students who wish to build their mathematical strength in order to excel academically and eventually win in Math Competitions (AMC, AIME, USAMO, IMO, ARML, MathCounts, Math Olympiad, ZIML, and other math leagues and tournaments, etc.), Science Fairs (County Science Fairs, State Science Fairs, national programs like Intel Science and Engineering Fair, etc.) and Science Olympiads, or for students who purely want to enrich their academic lives by taking more challenging courses and developing outstanding analytical, logical, and creative problem solving skills.

Since 2004 Areteem Institute has been teaching with methodology that is highly promoted by the new Common Core State Standards: stressing the conceptual level understanding of the math concepts, problem solving techniques, and solving problems with real world applications. With the guidance from experienced and passionate professors, students are motivated to explore concepts deeper by identifying an interesting problem, researching it, analyzing it, and using a critical thinking approach to come up with multiple solutions.

Thousands of math students who have been trained at Areteem have achieved top honors and earned top awards in major national and international math competitions, including Gold Medalists in the International Math Olympiad (IMO), top winners and qualifiers at the USA Math Olympiad (USAMO/JMO) and AIME, top winners at the

Zoom International Math League (ZIML), and top winners at the MathCounts National Competition. Many Areteem Alumni have graduated from high school and gone on to enter their dream colleges such as MIT, Cal Tech, Harvard, Stanford, Yale, Princeton, U Penn, Harvey Mudd College, UC Berkeley, or UCLA. Those who have graduated from colleges are now playing important roles in their fields of endeavor.

Further information about Areteem Institute, as well as updates and errata of this book, can be found online at http://www.areteem.org.

Acknowledgments

This book contains many years of collaborative work by the staff of Areteem Institute. This book could not have existed without their efforts. Huge thanks go to the Areteem staff for their contributions!

The examples and problems in this book were either created by the Areteem staff or adapted from various sources, including other books and online resources. Especially, some good problems from previous math competitions and contests such as AMC, AIME, ARML, MATHCOUNTS, and ZIML are chosen as examples to illustrate concepts or problem-solving techniques. The original resources are credited whenever possible. However, it is not practical to list all such resources. We extend our gratitude to the original authors of all these resources.

Copyright © ARETEEM INSTITUTE. All rights reserved.

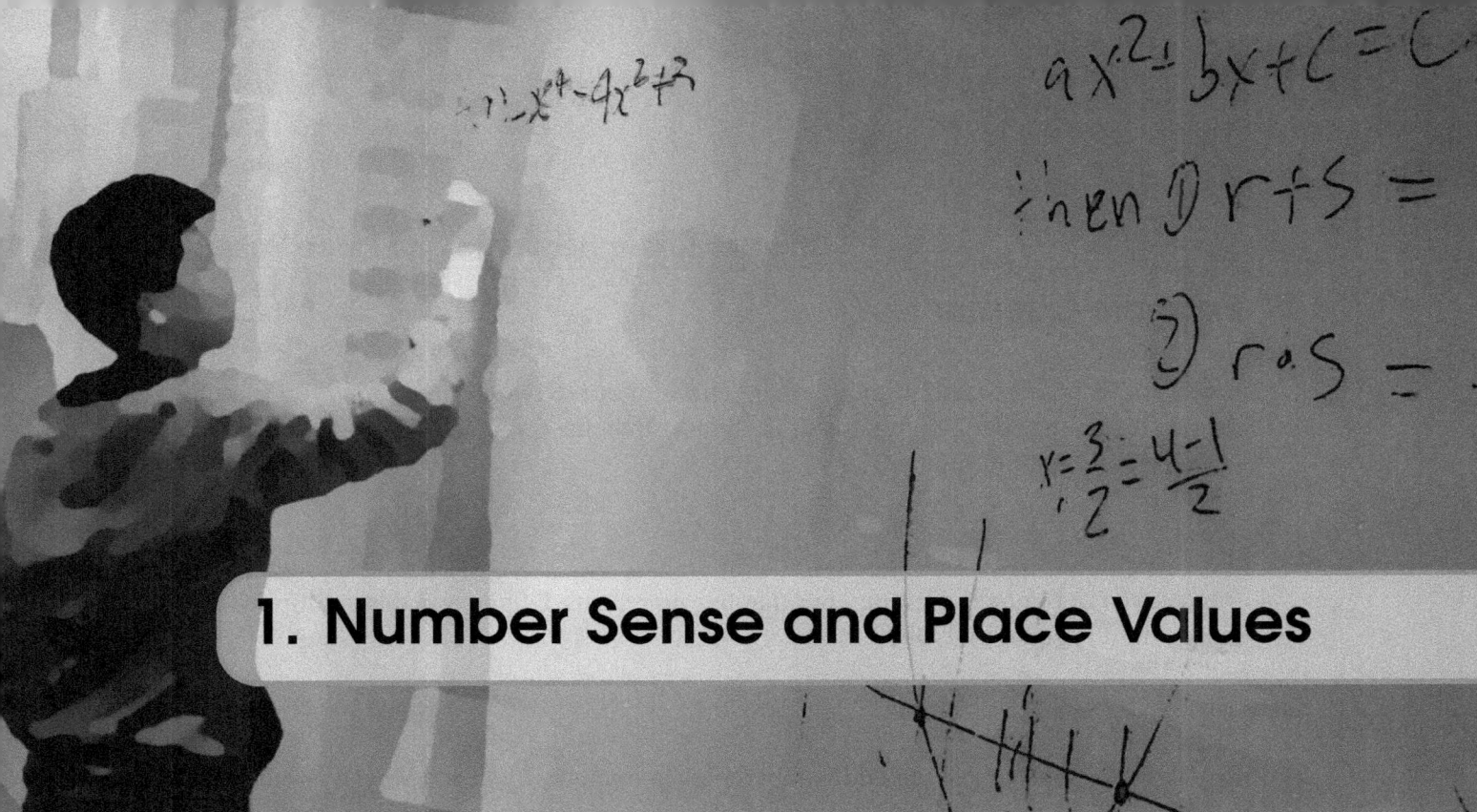

1. Number Sense and Place Values

Number Theory Introduction

- The theory of numbers goes back 3,000 years or more to the time of the Babylonians.
- The oriental cultures, Hindus and Chinese, also independently developed theories regarding numbers and their properties.
- Number theory, along with geometry, can probably be regarded as the first serious exploration into mathematics.
- Today many problems in math competitions have their roots in number theory.
- More specifically, "number" in Number Theory generally means *whole numbers* or *natural numbers* which include $0, 1, 2, 3, \ldots$. The set of these numbers is often denoted $\mathbb{N}$.
- Sometimes we will also talk about integers $\ldots, -2, -1, 0, 1, 2, \ldots$. The set of these numbers is often denoted $\mathbb{Z}$.

Place Values

- The value of a digit depends on its place, or position in a number.
- We call the digits (in reverse order) the units/ones digit, the tens digit, the hundreds digit, etc.
- For example, $654 = 6 \times 100 + 5 \times 10 + 4 \times 1$. That is, the number 654 has hundreds digit 6, tens digit 5, and units digit 4.
- Similarly, $2016 = 2 \times 1000 + 0 \times 100 + 1 \times 10 + 6 \times 1$.

Copyright © ARETEEM INSTITUTE. All rights reserved.

1.1 Example Questions

Problem 1.1 There is a 2-digit number. The sum of the two digits is 8. If you switch the two digits, the new 2-digit number is 18 more than the original 2-digit number. What is the original 2-digit number?

Problem 1.2 Find a 6-digit number that has no repeated digits and that multiplied by its last digit is equal to 999999.

Problem 1.3 Do you get an odd number or even number if you add 9 even numbers?

Problem 1.4 Uncle Jim got lost while we were driving back to California from Montréal. He saw a sign that said how many miles away we were from Los Angeles. Since he was driving fast, he couldn't quite see the number, but he knew it had 4 digits. I saw the number, but I wanted to have some fun so instead of telling him the number right away I gave him some clues:

- The number has the digit 1 somewhere.
- The digit in the hundreds' place is three times the digit in the thousands' place.
- The digit in the ones' place is 4 times the digit in the tens' place.
- The thousands' digit is 2.

How far away are we from Los Angeles?

Problem 1.5 Do the following calculations:

(a) $92 + 86 + 91 + 88 + 87 + 90 + 89 + 93 + 92 + 88$.

1.1 Example Questions

(b) $54 + 67 + 33 + 84 + 46 + 64 + 16$.

Problem 1.6 In the following calculations try to simplify before multiplying and dividing.
(a) $213 \times 24 \div 12$

(b) $75 \times 32 \div 400$

(c) $36 \times 62 \div 2 \times 7 \div 3$

Problem 1.7 What do we get if we multiply each of the following numbers by 11?

(a) 253

(b) 3594

(c) 45729

Problem 1.8 Square numbers that end in 5
(a) 15^2

(b) 165^2

Copyright © ARETEEM INSTITUTE. All rights reserved.

(c) 2005^2

Problem 1.9 Multiply numbers that start with the same digits and end in digits that add up to 10.
(a) 48×42

(b) 39×31

(c) 264×266

Problem 1.10 Troy bought some candies at the local grocery store in town. He had to pay $6.23, so he gave the cashier a $10 bill. The cashier gave him back three $1 bills and... 77 pennies! It seems there were no other coins available in the cash register. No one likes pennies and neither does Troy. He looked in his pockets for some coins that he could use instead, and he found 1 dime, 3 quarters, 2 pennies and 2 nickels. What coins will help him get the smallest amount of pennies back?

1.2 Quick Response Questions

Problem 1.11 What number has 12 hundreds and 9 ones?

Problem 1.12 What number has 65 ones and 3 thousands?

Problem 1.13 What 3-digit number has 8 hundreds, 1 more ten than hundreds, and no ones?

Problem 1.14 What is the difference between the largest 5-digit number and the smallest 5-digit number?

Problem 1.15 Calculate
$$47 + 56 + 43 + 60 + 43 + 52 + 40 + 48 + 45 + 58 + 41 + 55.$$

Problem 1.16 Calculate $480 \times 12 \div 800 \times 25$.

Problem 1.17 What is 43×11?

Problem 1.18 What is 63×11?

Problem 1.19 What is 95^2?

Problem 1.20 What is 23×27?

1.3 Practice Questions

Problem 1.21 What 3-digit numbers number have 1 ten and 1 more one than tens?

Problem 1.22 Fill in the blanks using place values:

$$\underline{\hspace{1cm}} + 700 + \underline{\hspace{1cm}} + 20 = 9728$$

Problem 1.23 Can you find three numbers whose sum is 100 and such that only one of the numbers is odd? Why or why not?

Problem 1.24 George's family lives in a house with a four-digit street number. The difference of the first digit and the last digit is 8. The 2nd digit is twice the first digit, and the 3rd digit is twice the 2nd digit. What is the street number of George's house?

Problem 1.25 Add the following numbers. Try to pair numbers to make the calculation quick.

(a) $191 + 809 + 259 + 2329 + 1741$

(b) $829 + 571 - 692 - 308$

Problem 1.26 Before multiplying and dividing, make sure to simplify some of the numbers in there.

(a) $350 \times 22 \div 7 \div 2$

(b) $715 \div 22 \times 4$

Problem 1.27 Multiply the following numbers by 11

(a) 743

(b) 3241

Problem 1.28 Calculate the following squares.

(a) 305^2

(b) 815^2

Problem 1.29 Find the product of the following pairs of numbers. Notice that they start with the same digits and their last digits add up to 10.

(a) 67×63

(b) 783×787

1.3 Practice Questions

Problem 1.30 Leslie just broke her piggy bank and is trying to figure out how much money she has. She counted the number of coins of each kind that she has: 354 pennies, 44 nickels, 12 dimes 72 quarters, 46 half dollars and 5 silver dollars. How much money did Leslie have in her piggy bank?

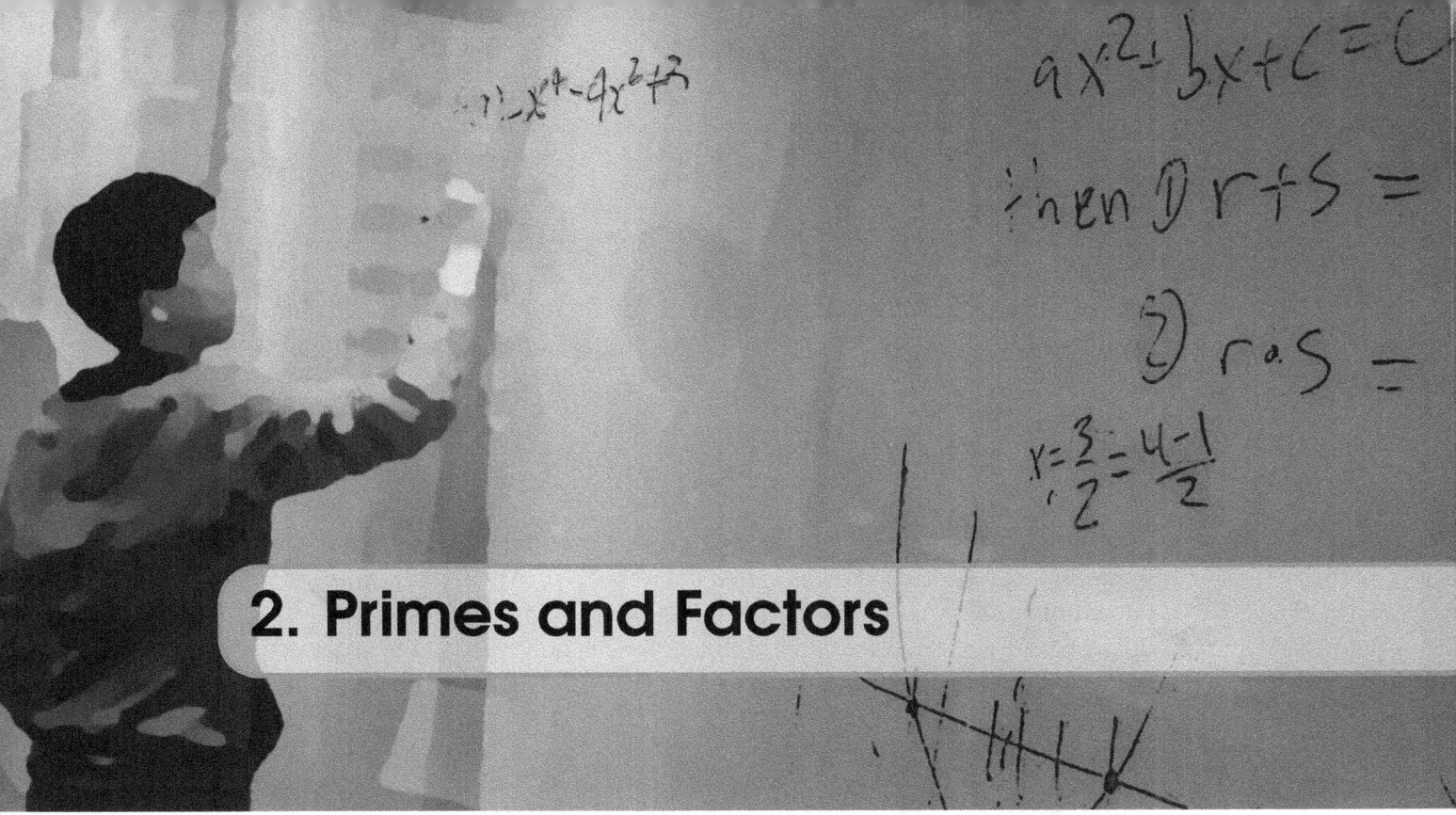

2. Primes and Factors

Divisibility

- If one number a goes evenly into another number b, we say b is divisible by a, denoted $a \mid b$. We say that a is a *divisor* or *factor* of b
- Equivalently, if $a \mid b$ then there is an integer k such that $b = a \cdot k$. In this case we say b is a multiple of a.
- For example, because $6 \cdot 5 = 30$ we have $6 \mid 30$ or 30 is divisible by 6 or 6 is a divisor/factor or 30. We also would say that 30 is a multiple of 6. Note: clearly $5 \mid 30$, etc. as well.
- **Caution**: The number 0 is a multiple of every integer (including itself). However, we cannot divide by 0, so 0 is never a factor of another number.

Prime Numbers

- The fundamental building block of number theory is the prime numbers.
- A natural number is *prime* if it has exactly two divisors. In other words, a prime number is divisible only by 1 and itself.
- A natural number is *composite* if it can be written as a product of two smaller numbers that are both greater than 1 (recall that by number we mean integer!). That is, a number N is composite if we can write $N = a \cdot b$ for $a, b > 1$.
- The natural numbers 0 and 1 are neither prime nor composite, but every other natural number is either prime or composite.
- The first few primes are $2, 3, 5, 7, \ldots$.

Copyright © ARETEEM INSTITUTE. All rights reserved.

- In fact, there are infinitely many primes! You do not need to know the proof for this class, but some hints are given as one of the challenge questions below.

Prime Factorization

- If a number is prime, then clearly we can write it as a product of primes (just itself!).
- In general, every number can be written as a product of primes (possibly with some primes repeated). This is called a number's *prime factorization*.
- For example, $12 = 2 \cdot 2 \cdot 3 = 2^2 \cdot 3$ and $300 = 2^2 \cdot 3 \cdot 5^2$.
- In fact, the prime factorization is unique, there is only one way to write a number as a product of primes (where we do *not* care about the order of the primes).

2.1 Example Questions

Problem 2.1 (Checking Whether a Number is Prime)

(a) Is the number 37 prime? How can you prove this?

(b) Expand your method in part (a) to work for any integer N. Explain why it works!

Problem 2.2 (Sieve of Eratosthenes) One of the earliest methods for finding prime numbers is called the *sieve of Eratosthenes*. It works like this: Start with a table of numbers; Cross out the number 1; Keep the number 2 as a prime, and cross out all other multiples of 2 (the even numbers); Keep the next number, 3, as a prime, and cross out all other multiples of 3; Keep the next number, 5, as a prime, and cross out all other multiples of 5; and so on. At the end, all remaining numbers on the table are primes.

2.1 Example Questions

(a) Using the table below, find all the prime numbers less than 180.

1	2	3	4	5	6	7	8	9	10	11	12
13	14	15	16	17	18	19	20	21	22	23	24
25	26	27	28	29	30	31	32	33	34	35	36
37	38	39	40	41	42	43	44	45	46	47	48
49	50	51	52	53	54	55	56	57	58	59	60
61	62	63	64	65	66	67	68	69	70	71	72
73	74	75	76	77	78	79	80	81	82	83	84
85	86	87	88	89	90	91	92	93	94	95	96
97	98	99	100	101	102	103	104	105	106	107	108
109	110	111	112	113	114	115	116	117	118	119	120
121	122	123	124	125	126	127	128	129	130	131	132
133	134	135	136	137	138	139	140	141	142	143	144
145	146	147	148	149	150	151	152	153	154	155	156
157	158	159	160	161	162	163	164	165	166	167	168
169	170	171	172	173	174	175	176	177	178	179	180

(b) Explain why the sieve of Eratosthenes works.

Problem 2.3 A "prime-prime" number is a prime number that yields a prime number when its units digit is omitted. For example, 131 is a three-digit "prime-prime" number because 131 is prime and 13 is prime.

(a) How many two-digit "prime-primes" are there?

(b) Without actually counting them, argue that there "should" be more "prime-prime" numbers between $100-200$ than between $200-300$.

Problem 2.4 For each of the following numbers, determine whether the number is prime or not. If the number is not prime, give an example of a non-trivial divisor.

(a) 457

(b) 301

(c) 657

(d) 247

Problem 2.5 Determine a value d such that $1d3$ and $5d7$ are prime numbers.

2.1 Example Questions

Problem 2.6 Consider the number 2016.

(a) Find the prime factorization of 2016.

(b) What is the sum of the *distinct* prime factors of 2016?

Problem 2.7 The numerical values of the years are favorite numbers of many math contests. The prime factorizations of these numbers are often useful in those contests. We already found the prime factorizations of 2016 in Problem 2.6 Find the prime factorizations of the following:

(a) 2017

(b) 2018

(c) 2019

(d) 2015

Problem 2.8 Let N be the largest seven-digit number that can be constructed using each of the digits $1, 2, 3, 4, 5, 6, 7$ such that the sum of every pair of consecutive digits is a prime number. What is the number N?

Problem 2.9 Consider the numbers $17, 19, 21, 25, 26$.

(a) Which of the numbers has the smallest prime factor?

(b) Which of the numbers has the largest prime factor?

Problem 2.10 Find the prime factorization of 510510.

2.2 Quick Response Questions

Problem 2.11 Find all even primes.

Problem 2.12 What is the smallest prime number greater than 50?

Problem 2.13 What is the largest prime number less than 100?

Problem 2.14 How many prime numbers are between 10 and 30?

Problem 2.15 Two primes are twin primes if one is 2 more than the other. For example, 3 and 5 are a pair of twin primes. How many twin primes pairs are less than 50?

Problem 2.16 Find the prime factorization of 60. What is the largest prime number in the prime factorization?

Problem 2.17 Find the prime factorization of 62. What is the largest prime number in the prime factorization?

Problem 2.18 A proper factor is a factor of a given number n not equal to n. What is the largest proper factor of 62?

Problem 2.19 What is the largest proper factor of 60?

Problem 2.20 How many factors of 63 are there?

2.3 Practice Questions

Problem 2.21 Is 181 a prime number?

Problem 2.22 Extend the sieve of Eratosthenes two more rows by finding all primes from 182 to 210. Hint: Multiples of 2 and 5 are easily eliminated. Now check for other prime factors of n up to $\sqrt{n}$.

Problem 2.23 Find all prime-prime numbers between 200 and 300.

Problem 2.24 Is 899 a prime number?

Problem 2.25 Find all palindromic primes less than 200, that is, all primes of the form $\overline{aa}$ or $\overline{aba}$.

Problem 2.26 Find the prime factorizations of 1776 and 2001.

Problem 2.27 Find the prime factorization of 1492 and 2525.

Problem 2.28 Let N be the largest seven-digit palindrome that can be constructed from the digits $1, 2, 3, 4, 5, 6, 7$ (repeats are allowed) such that the sum of every pair of consecutive digits is a prime number. What is the number N?

Problem 2.29 Find the common prime factor of 418, 437, 456.

Problem 2.30 Find the prime factorization of 969969.

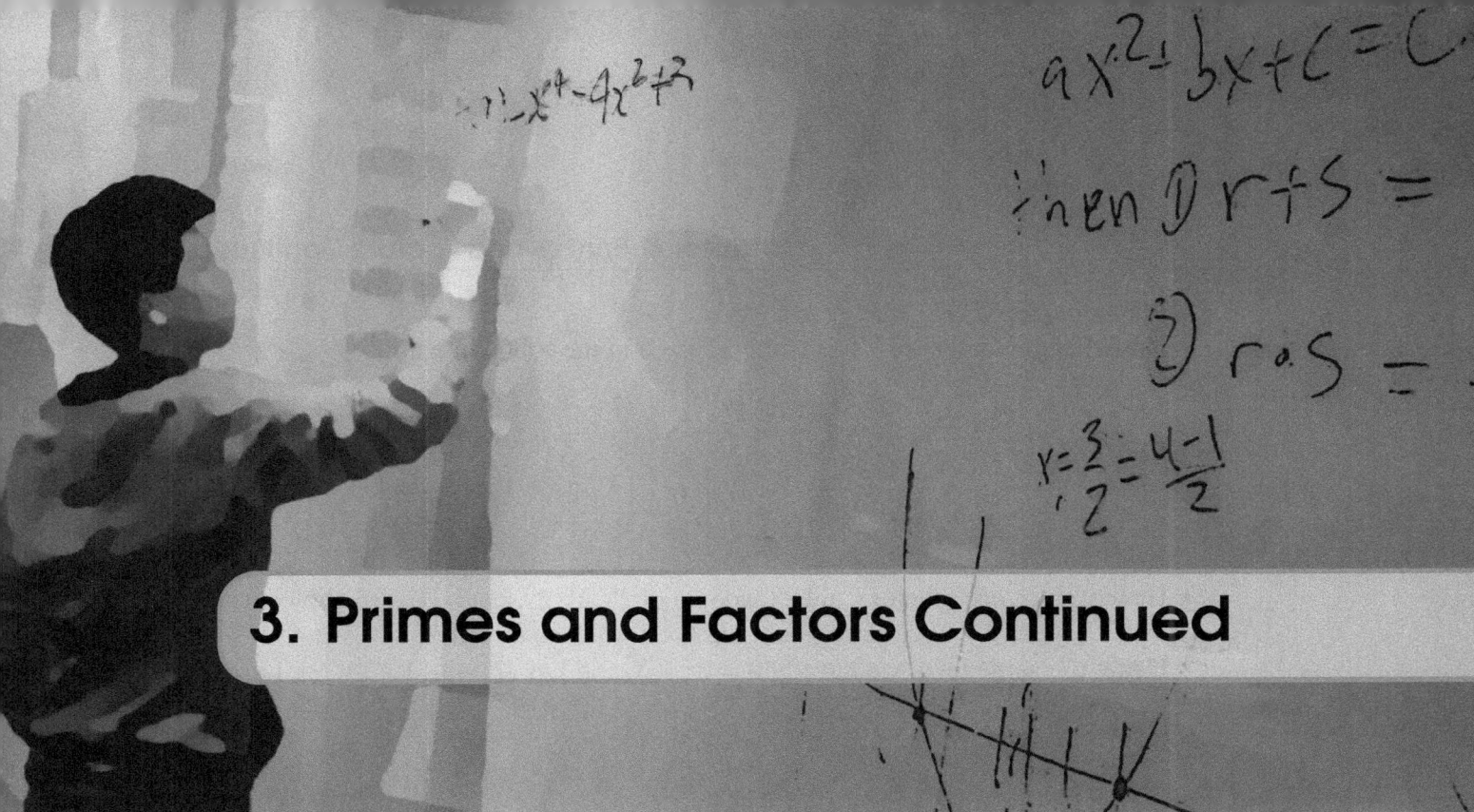

3. Primes and Factors Continued

Review of Divisibility

- Recall if one number a goes evenly into another number n, we say n is divisible by a, denoted $a \mid n$. We say that a is a *divisor* or *factor* of n.
- Equivalently, if $a \mid n$ then there is an integer b such that $n = a \cdot b$. In this case we say n is a multiple of a.
- Therefore, if $a \mid b$, with $n = a \cdot b$, then naturally $b \mid n$ as well. In this sense, the factors/divisors of n can be thought of as coming in "pairs".

Counting Multiples

- Half of all numbers are a multiple of 2 (the even numbers), so $50 = 100 \div 2$ of the numbers $1, 2, \ldots, 100$ are divisible by 2.
- 101 is an odd number, so there are still 50 multiples of 2 between 1 and 101. We can think of 50 here as $101 \div 2$ then rounded down.
- In general, if we want to find the number of multiples of k between 1 and n we do $n \div k$ and keep the quotient (and ignore the remainder). Alternatively, we can do $n \div k$ and then ignore any decimals (only keeping the whole number part).
- For example, since $99 \div 7 = 14$ with remainder 1, there are 14 multiples of 7 less than 100.

Prime Factorizations and Number of Factors

- Recall from last time that every number can be written (uniquely) as a product of prime numbers.

Copyright © ARETEEM INSTITUTE. All rights reserved.

- For example, $72 = 2 \times 2 \times 2 \times 3 \times 3 = 2^3 \times 3^2$.
- Note that the factors of 72 can be organized in the following table:

	2^0	2^1	2^2	2^3
3^0	1	2	4	8
3^1	3	6	12	24
3^2	9	18	36	72

- Hence, since each factor can be made up of $2^0, 2^1, 2^2, 2^3$ ($3+1 = 4$ choices) and $3^0, 3^1, 3^2$ ($2+1 = 3$ choices) there are $4 \times 3 = 12$ factors of 72.
- Using the same idea, we have $540 = 2^2 \times 3^3 \times 5$ has $(2+1)(3+1)(1+1) = 24$ factors.

3.1 Example Questions

Problem 3.1 Even Number of Factors?

(a) Use the "pairing" trick mentioned above to help find and count all the factors of 72.

(b) We might be tempted after part (a) to say that every number has an even number of factors. (Or maybe tempted to say every number greater than 1 has an even number of factors.) Disprove this hypothesis by using the pairing trick to help find and count all factors of 36.

(c) Come up with a hypothesis for when a number n has an even number of factors. You do not need to prove this now.

Problem 3.2 Of the first 50 positive integers, which have an odd number of factors and which have an even number of factors?

3.1 Example Questions

Problem 3.3 For any positive integer n, define $\boxed{n}$ to be the sum of the positive factors of n. For example, $\boxed{6} = 1 + 2 + 3 + 6 = 12$.

(a) Find $\boxed{20}$.

(b) Find $\boxed{\boxed{11}}$.

Problem 3.4 Counting Multiples Practice

(a) How many three-digit numbers are divisible by 17?

(b) How many whole numbers strictly between 1 and 2009 are divisible by 2 but not by 7?

Problem 3.5 Which number has more positive divisors, 2015 or 2020?

Problem 3.6 Find the smallest positive integer x such that

(a) $252 \cdot x$ is a perfect square.

(b) $252 \cdot x$ is a perfect cube.

Problem 3.7 A prison houses 1,000 inmates in 1,000 prison cells. It also has 1,000 guards. One day the guards decide to use a number theory problem to free some prisoners. The first guard goes through the prison and unlocks every cell. The second guard goes through the prison and locks every second cell. The third guard goes through the prison and changes the status of every third cell. That is, if it's locked, he unlocks it; and if it's unlocked, he locks it. This process continues (that is the nth prison guard changes the status of every nth cell) until the 1,000th guard has passed through the prison, after which every prisoner whose cell is left unlocked is free to leave. How many prisoners will be set free?

Problem 3.8 How many perfect squares are factors of 2592?

Problem 3.9 Find the smallest prime p such that $2^p - 1$ is not a prime number.

Problem 3.10 From the set of natural numbers $2, 3, 4, \ldots, 999$, delete 9 subsets as follows. First delete all even numbers except 2, then all multiples of 3, except 3, then all multiples of 5, except 5, and so on for the nine primes 2, 3, 5, 7, 11, 13, 17, 19, and 23. After all these steps, there are only three composite numbers left. Find the sum of these composite numbers.

3.2 Quick Response Questions

Problem 3.11 Find all positive integers n such that $n \mid 28$. What is the largest odd factor of 28?

Problem 3.12 Find the largest number less than 50 with an odd number of factors.

Problem 3.13 Find the sum of all divisors of 36.

Problem 3.14 How many positive multiples of 11 less than 111 are there?

Problem 3.15 How many multiples of 11 are there between 100 and 1000?

Problem 3.16 How many multiples of 11 less than 1000 are also multiples of 13?

Problem 3.17 Find the number of factors of 175.

Problem 3.18 Find the sum of the factors of 8.

Problem 3.19 Find the product of the factors of 8.

Problem 3.20 Note the number $9216 = 2^{10} \times 3^2$. How many factors does 9216 have?

3.3 Practice Questions

Problem 3.21 Prove that a number n has an odd number of factors if and only if it is a perfect square. That is, prove: (i) if n is a perfect square ($n = k^2$ for an integer k), then n has an odd number of factors and (ii) if n is not a perfect square then n has an even number of factors.

Problem 3.22 How many numbers between 99 and 500 have an odd number of factors?

Problem 3.23 Recall we defined $\boxed{n}$ to be the sum of the positive factors of n. Find $\boxed{\boxed{11}}$.

Problem 3.24 How many positive integers less than 2016 are divisible by 3 but not by 4?

Problem 3.25 Find the number of divisors for each of the numbers $6, 30, 210, 2310$. Do you see a pattern?

Problem 3.26 Find the smallest positive integer x such that $252 \cdot x$ is a perfect square and a perfect cube.

Problem 3.27 A new prison has 100 inmates in 100 prison cells, with 100 guards. All the cells start unlocked. The prison has an electronic locking system that needs the authorization of at least 4 guards to lock the door to a cell. The first guard goes through the prison and authorizes every cell to lock. Then the second guard goes through the prison and authorizes every second cell to lock. The rest of the guards continue in this manner (3rd guard authorizes every third cell, etc.) The process continues for all 100 guards, after which every prisoner whose cell is left unlocked is free to leave. How many prisoners will be set free?

Problem 3.28 How many perfect squares are factors of 11025?

Problem 3.29 A (positive) divisor a of $n > 0$ is proper if $a < n$. Find the sum of proper divisors of 496.

Problem 3.30 From the set of natural numbers $2, 3, 4, \ldots, 1999$, delete all even numbers except 2, then all multiples of 3 except 3, etc. until all the composite numbers are removed. How many prime numbers does it take to do this?

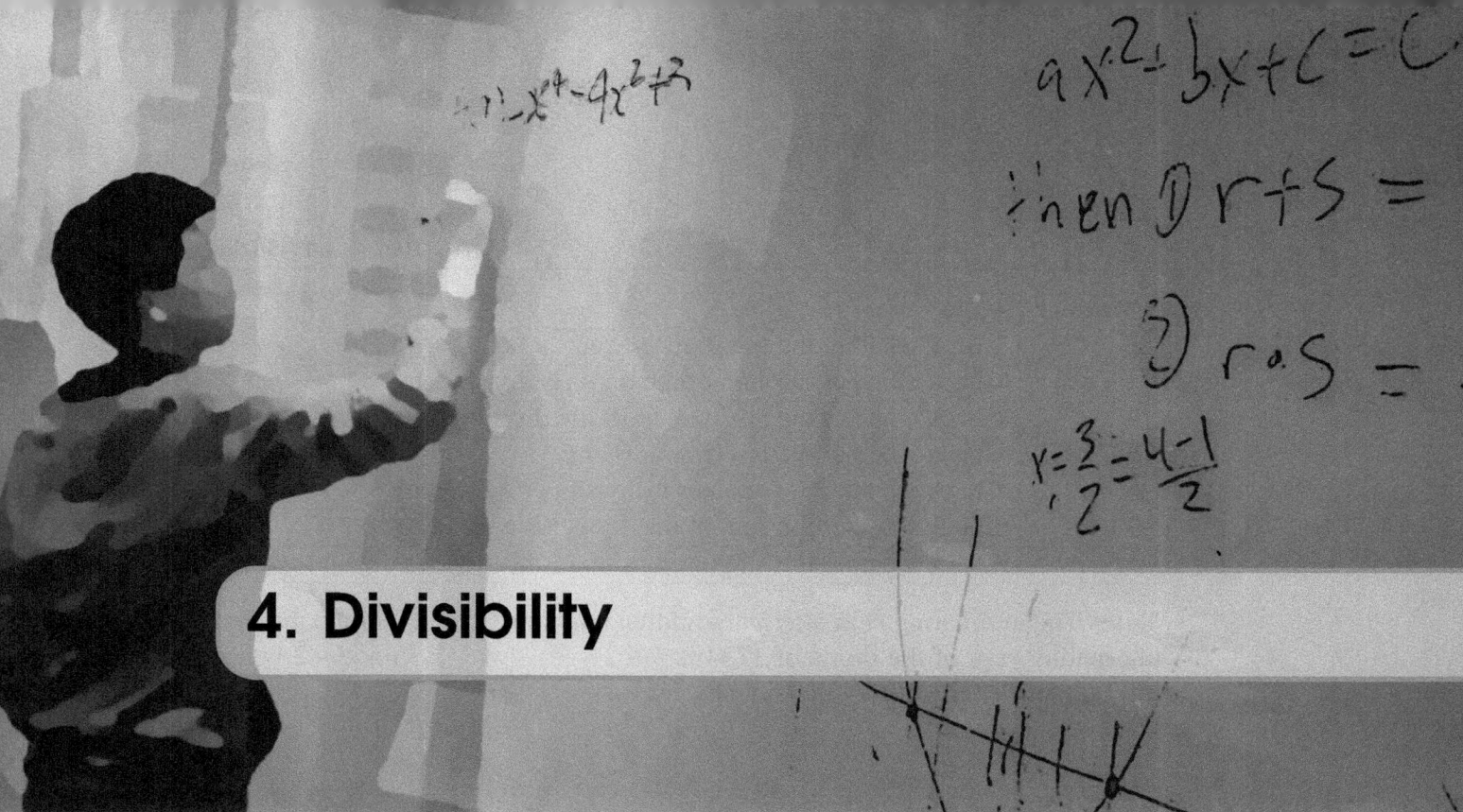

4. Divisibility

Review of Divisibility

- Recall if one number a goes evenly into another number n, we say n is divisible by a, denoted $a \mid n$. We say that a is a *divisor* or *factor* of n.
- Equivalently, if $a \mid n$ then there is an integer b such that $n = a \cdot b$. In this case we say n is a multiple of a.
- For small numbers, it is often helpful to have a "trick" or "rule" to help decide when another number is divisible by it. These are often called *divisibility rules*.

Review of Place Values

- The value of a digit depends on its place, or position in a number.
- We call the digits (in reverse order) the units/ones digit, the tens digit, the hundreds digit, etc.
- For example, $654 = 6 \times 100 + 5 \times 10 + 4 \times 1$. That is, the number 654 has hundreds digit 6, tens digit 5, and units digit 4.
- Similarly, $2016 = 2 \times 1000 + 0 \times 100 + 1 \times 10 + 6 \times 1$.

Divisibility Rules

- The most common divisibility rules are summarized below:

Number	Rule
2	Check whether the last digit is even.
3	Check whether the sum of the digits is divisible by 3.
4	Check whether the last two digits are divisible by 4.
5	Check whether the last digit is 0 or 5.
6	Check whether the number is divisible by 2 and by 3.
9	Check whether the sum of the digits is divisible by 9.
11	Check whether the alternating sum of the digits is divisible by 11.

- Note: The alternating sum alternates adding and subtracting. For example, the alternating sum of the digits of 1234 is $1-2+3-4 = -2$. Since -2 is not a multiple of 11, 1234 is not divisible by 11.

4.1 Example Questions

Problem 4.1 Come up with a divisibility rule for the following numbers. (That is, find an easy way to check if a number is divisible by the following numbers.) Can you explain why the rules work?

(a) 2.

(b) 5.

Problem 4.2 Divisibility Rule for Three

(a) Write out the representation of the number 15624 in terms of its digits and place values.

(b) Show that "Is 15624 divisible by 3?" has the same answer as "Is $1+5+6+2+4$ divisible by 3?". Hint: What is 9×1? 9×11? 9×111? 9×1111?

4.1 Example Questions

(c) Use part (b) to state a general divisibility rule for 3.

Problem 4.3 Come up with divisibility rules for the following numbers.

(a) 6

(b) 4

Problem 4.4 What five-digit multiple of 11 consists entirely of 2s and 3s?

Problem 4.5 A number $\overline{35a2a}$ is divisible by 3. Its last three digits form a three-digit number $\overline{a2a}$ that can be exactly divided by 2. Find this five-digit number.

Problem 4.6 A four digit number $\overline{7a4b}$ is divisible by 18. Find the value of a and b so that this four-digit number has the least value.

Problem 4.7 Find all possible six-digit numbers represented by 2016__ __ which can exactly be divided by 2, 3, 4, 5.

Problem 4.8 For how many positive integer values of n are both $\dfrac{n}{3}$ and $3n$ three-digit integers?

Problem 4.9 How many three-digit numbers are divisible by 13?

Problem 4.10 A natural number is a multiple of both 3 and 4, and has totally 10 divisors. What is this number?

4.2 Quick Response Questions

Problem 4.11 Find the sum of divisors of 98.

Problem 4.12 Find the number of factors of 360.

Problem 4.13 Find the number of factors of 980?

Problem 4.14 Is 980 a perfect square?

Problem 4.15 Is the number 5201928 divisible by 9?

Problem 4.16 Is the number 5201928 divisible by 11?

Problem 4.17 What is the smallest number, made up of only 4's (so $4, 44, 444, 4444, \ldots$), that is divisible by 3?

Problem 4.18 How many of the following numbers are divisors of 330: $2, 3, 4, 5, 6, 9, 11$?

Problem 4.19 How many of the following numbers are divisors of 2134: $2, 3, 4, 5, 6, 9, 11$?

Problem 4.20 What is the smallest number that passes all of the divisibility tests at the beginning of this week's handout? That is, what is the smallest number that is divisible by 2, 3, 4, 5, 9, and 11.

4.3 Practice Questions

Problem 4.21 Come up with a divisibility rule for:

(a) 10

(b) Compare to the divisibility rules for 2 and 5, do you see a connection?

Problem 4.22 Show the number 55215 is divisible by 9. Justify your answer by writing out the representation of 55215 in terms of its digits and place values to show that 55215 is divisible by 9 if and only if the sum of its digits is divisible by 9.

Problem 4.23 Come up with divisibility rules for the following numbers.

(a) 12

(b) 8

Problem 4.24 Find a seven-digit multiple of 11 consisting of entirely 2s and 3s. Is there only one?

Problem 4.25 A number $\overline{35a2a}$ is divisible by 2. Its last three digits form a three-digit number $\overline{a2a}$ that can be exactly divided by 3. Find all possible such five-digit numbers.

Problem 4.26 A four digit number $\overline{7a4b}$ is divisible by 18. Find the value of a and b so that this four-digit number has the largest value.

Problem 4.27 How many six-digit numbers represented by 2016__ __ can be exactly divided by 3 and 5?

Problem 4.28 For how many positive integer values of n are both $n/3$ and $n/9$ two-digit integers.

Problem 4.29 How many numbers less than 1000 are there that are divisible by 11 but not 11^2?

Problem 4.30 Find all numbers that are multiples of 2, 3, 5 and have 12 divisors.

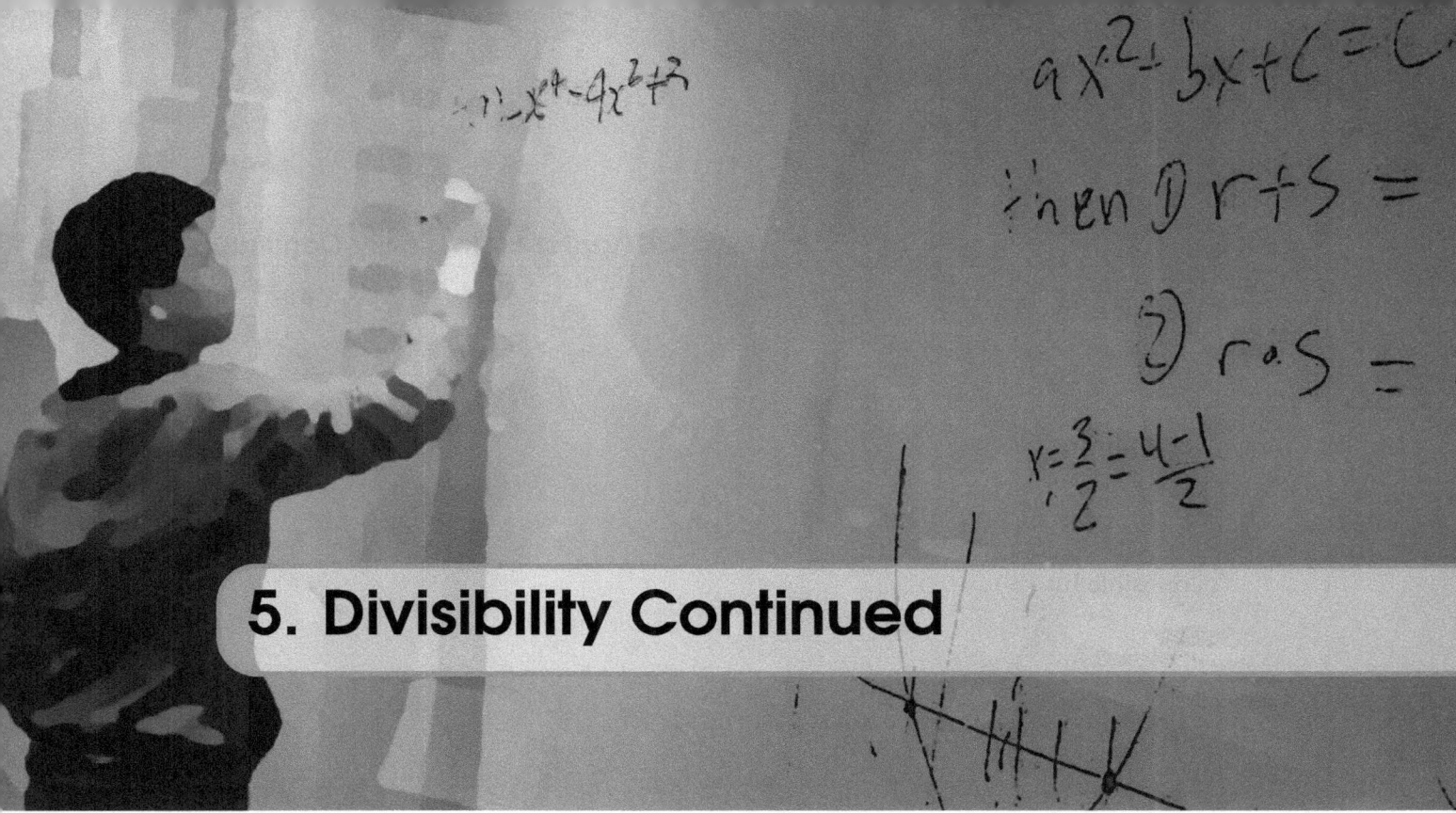

5. Divisibility Continued

Divisibility Rules Review

- Recall the common divisibility rules:

Number	Rule
2	Check whether the last digit is even.
3	Check whether the sum of the digits is divisible by 3.
4	Check whether the last two digits are divisible by 4.
5	Check whether the last digit is 0 or 5.
6	Check whether the number is divisible by 2 and by 3.
9	Check whether the sum of the digits is divisible by 9.
11	Check whether the alternating sum of the digits is divisible by 11.

- Recall: The alternating sum alternates adding and subtracting. For example, the alternating sum of the digits of 1234 is $1-2+3-4 = -2$. Since -2 is not a multiple of 11, 1234 is not divisible by 11.

5.1 Example Questions

Problem 5.1 What digit could fill in the blank to make 89__43

Copyright © ARETEEM INSTITUTE. All rights reserved.

(a) divisible by 11?

(b) divisible by 9?

(c) divisible by 4?

Problem 5.2 What is the smallest positive integer,

(a) that is divisible by 2 and 3 that consists entirely of 2's and 3's, with at least one of each?

(b) entirely consisting of 4 and 9, at least one of each, that is divisible by both 4 and 9?

Problem 5.3 How many positive multiples of 3 less than 1000 use only the digits 2 and/or 4?

Problem 5.4 A four-digit number $\overline{7x4y}$ should be exactly divisible by 55. Find all possible four-digit numbers that meet this requirement.

Problem 5.5 The number 64 has the property that it is divisible by its units digit. How many whole numbers between 10 and 50 have this property?

5.1 Example Questions

Problem 5.6 A positive 8-digit integer has only 2 different digits. What is the smallest such number that is a multiple of both 5 and 6?

Problem 5.7 There are four pupils whose ages are four consecutive integers. The product of their ages is 5040. What are their ages respectively?

Problem 5.8 Barry wrote 6 different numbers, one on each side of 3 cards, and laid the cards on a table, as shown. The sums of the two numbers on each of the three cards are equal. The three numbers on the hidden sides are prime numbers. What is the average of the hidden prime numbers?

Problem 5.9 A five-digit number consists of distinct digits. It is divisible by 3, 5, 7, and 11. Find the greatest five-digit number that satisfies these conditions.

Problem 5.10 In the multiplication problem below, A, B, C and D are distinct digits. What is $A+B$?

$$\begin{array}{r} A\ B\ A \\ \times\qquad C\ D \\ \hline C\ D\ C\ D \end{array}$$

5.2 Quick Response Questions

Problem 5.11 What is the largest prime number in the prime factorization of 1200?

Problem 5.12 How many factors does 1248 have?

Problem 5.13 Is 117 a prime number?

Problem 5.14 List all numbers between 105 and 125 that are divisible by 4. How many numbers are in the list?

Problem 5.15 Is 3474 divisible by 9?

Problem 5.16 Is 3474 divisible by 11?

Problem 5.17 Is 1527 is divisible by 3, but not by 9?

Problem 5.18 Is 1527 a perfect square?

Problem 5.19 What is the largest prime factor of 10035?

5.2 Quick Response Questions

Problem 5.20 What is the largest prime factor of 17116?

5.3 Practice Questions

Problem 5.21 For what values of x is $1x273x95$

(a) divisible by 9?

(b) divisible by 11?

Problem 5.22 Find the smallest positive integer consisting entirely of 3's and 5's, at least one of each, that is divisible by 3 and 5.

Problem 5.23 Find all multiples of 9 less than 10000 whose only digits are 1 and/or 8?

Problem 5.24 A four-digit number $\overline{2x5y}$ should be exactly divisible by 36. Find all possible four-digit numbers that meet this requirement.

Problem 5.25 The number 24 has the property that it is divisible by its lead digit. How many whole numbers (strictly) between 20 and 60 have this property?

Problem 5.26 Find all six digit numbers consisting entirely of 0's and 1's that are divisible by 6.

5.3 Practice Questions

Problem 5.27 The product of two 2-digit natural numbers is 6975. What is the smaller number?

Problem 5.28 Cary wrote the six numbers 4, 5, 6, 7, 9, and 10, one on each side of 3 cards. The sum of the two numbers on each card is a different prime number. How are the numbers paired up on the cards?

Problem 5.29 What is the smallest five-digit integer divisible both by 8 and by 9?

Problem 5.30 In the addition problem below, A, B, C and D are distinct digits. What is $C + D$?

$$\begin{array}{r} A\ B\ A \\ +\ \ \ \ C\ D \\ \hline B\ A\ B \end{array}$$

6. GCD's and LCM's

Common Divisors

- Suppose m and n are two positive integers. If another positive integer k is a divisor of both m and n, then k is a *common divisor* or *common factor* of m and n.
- For example, 3 is a common divisor of 12 and 18.
- The number 1 is a common divisor of all positive integers.
- For two particular numbers, there are sometimes more than one common divisors.
- For example, the common divisors of 12 and 18 include $1, 2, 3$, and 6. Among them, 6 is the greatest.

Greatest Common Divisor (GCD)

- The *greatest common divisor (GCD)* of two numbers m and n is denoted $\gcd(m, n)$.
- For example, we have $\gcd(12, 18) = 6$.
- The greatest common divisor can be found by looking at the prime factorization of the two numbers involved. Using the previous example, $12 = 2^2 \times 3$, and $18 = 2 \times 3^2$. For each of the primes 2 and 3, we take the smaller exponents (which are both 1), and get $2 \times 3 = 6$. Therefore $\gcd(12, 18) = 6$.
- If two numbers m and n don't have any common prime factors, we have $\gcd(m, n) = 1$, and the numbers m and n are said to be *relatively prime*, or *co-prime*.
- For example, $\gcd(8, 15) = 1$, so 8 and 15 are relatively prime.
- The concept of GCD can naturally be extended to more than two numbers. For example, $\gcd(24, 36, 48) = 12$.
- One simple application using the GCD is reducing fractions: If the numerator and

Copyright © ARETEEM INSTITUTE. All rights reserved.

denominator of a fraction are not relatively prime, we can divide both of them by their GCD: $\dfrac{12}{18} = \dfrac{12 \div 6}{18 \div 6} = \dfrac{2}{3}$.

Common Multiples

- Suppose m and n are two positive integers. If another positive integer k is a multiple of both m and n, then k is a *common multiple* of m and n.
- For example, 180 is a common multiple of 12 and 18.
- For two particular numbers, there are infinitely many common multiples.
- For example, the common multiples of 12 and 18 include $36, 72, 108, 144$, and so on. Among them, 36 is the smallest.

Least Common Multiple (LCM)

- The *least common multiple (LCM)* of two numbers m and n is denoted $\text{lcm}(m,n)$.
- For example, $\text{lcm}(12,18) = 36$.
- The least common multiple can be found by looking at the prime factorization of the two numbers involved. Using the previous example, $12 = 2^2 \times 3$, and $18 = 2 \times 3^2$. For each of the primes 2 and 3, we take the larger exponents (which are both 2 in this case), and get $2^2 \times 3^2 = 36$. Therefore $\text{lcm}(12,18) = 36$.
- The concept of LCM can be extended to more than two numbers. For example, $\text{lcm}(12,18,24) = 72$, where 72 is the least multiple for all three numbers 12, 18, and 24.

6.1 Example Questions

Problem 6.1 Find the greatest common divisor and the least common multiple of

(a) 123 and 1681.

(b) 8! and 4^3 (Recall 8! represents "8 factorial", which means $1 \times 2 \times 3 \times \cdots 8$.)

6.1 Example Questions

(c) $3! + 5!$ and $5! + 6!$.

Problem 6.2 What is the least common multiple of the first 10 positive integers?

Problem 6.3 Among the fractions $\dfrac{1}{24}, \dfrac{2}{24}, \dfrac{3}{24}, \ldots, \dfrac{23}{24}$, how many are irreducible? (Irreducible means the numerator and denominator are relatively prime)

Problem 6.4 The least common multiple of two numbers is 180 and their greatest common divisor is 30. One of the two numbers is 90. what is the other number?

Problem 6.5 True or False. You do not need to formally prove your answers, but given an example with explanation for each.

(a) If m, n are relatively prime and n, l are relatively prime then m, l are relatively prime.

(b) Any common divisors of m and n are divisors of $\gcd(m, n)$.

(c) Any common multiples of m and n are multiples of $\text{lcm}(m, n)$.

Problem 6.6 A certain fraction, $\dfrac{m}{n}$, if reduced to lowest terms, equals $\dfrac{5}{11}$. Also given that $m + n = 80$. What are m and n?

Copyright © ARETEEM INSTITUTE. All rights reserved.

Chapter 6. GCD's and LCM's

Problem 6.7 How many integers between 1000 and 2000 have all three of the numbers 12, 18 and 30 as factors? What are they?

Problem 6.8 How many positive two-digit integers are factors of both 360 and 540? What are they?

Problem 6.9 Three units commonly used to measure angles are degrees (360 degrees in a circle), grads (400 grads in a circle) and mils (6400 mils in a circle). A right angle has an integer value for all three units which is 90 degrees, 100 grads and 1600 mils. Find the number of degrees of the smallest positive angle which is an integer for all three units.

Problem 6.10 Adding the same positive integer N to both the numerator and denominator of the fraction $\frac{1997}{2000}$, we obtain a new fraction that is equal to $\frac{2000}{2001}$. What is this number N?

6.2 Quick Response Questions

Problem 6.11 How many factors does 630 have?

Problem 6.12 Find the prime factorizations of 124 and 741. What is the sum of the largest prime numbers in the prime factorizations?

Problem 6.13 Are 124 and 741 relatively prime?

Problem 6.14 Find the GCD of 105 and 385.

Problem 6.15 Find the GCD of 630 and 385.

Problem 6.16 Find the LCM of 105 and 385.

Problem 6.17 Find the GCD of 525 and 385.

Problem 6.18 Find the LCM of 126 and 165.

Problem 6.19 Find all common divisors of 20 and 30. What is the GCD of 20 and 30?

Problem 6.20 Find all common multiples of 20 and 30 that less than 600. What is the LCM of 20 and 30?

6.3 Practice Questions

Problem 6.21 Find the GCD and LCM of $2!+3!+4!$ and $3!+4!+5!$.

Problem 6.22 What is the least common multiple of 11, 12, 13, 14 and 15?

Problem 6.23 Which of the fractions $\frac{1}{9}, \frac{2}{9}, \frac{3}{9}, \ldots, \frac{8}{9}$ are in irreducible form?

Problem 6.24 The least common multiple of two numbers is 468 and their greatest common divisor is 4. One of the two numbers is 36. what is the other number?

Problem 6.25 Suppose you have two integers m, n.

(a) If m, n are relatively prime, find $\text{lcm}(m, n)$.

(b) If instead $m \mid n$, find $\gcd(m, n)$ and $\text{lcm}(m, n)$.

Problem 6.26 A certain fraction, $\frac{m}{n}$, if reduced to lowest terms, equals $\frac{13}{17}$. Also given that $n - m = 76$. What are m and n?

Problem 6.27 How many integers strictly between 1000 and 2000 have all of the numbers 6, 14, 15, and 35 as factors? What are they?

Problem 6.28 Find all factors of lcm(90, 108) that are divisible by gcd(90, 108).

Problem 6.29 Find the smallest positive integer k such that $\dfrac{45 \cdot k}{117}$ and $\dfrac{49 \cdot k}{91}$ are both whole numbers.

Problem 6.30 Find a positive integer N such that, when it is added to both the numerator and denominator of the fraction $\dfrac{39}{129}$, we obtain a new fraction equal to $\dfrac{23}{29}$.

7. GCD's and LCM's Continued

Review of GCD's and LCM's

- The *greatest common divisor* (GCD) of m and n (denoted $\gcd(m,n)$) is the largest number d such that $d \mid m$ and $d \mid n$.
- We call two numbers m, n *relatively prime* if $\gcd(m,n) = 1$.
- The *least common multiple* (LCM) of m and n (denoted $\text{lcm}(m,n)$) is the smallest number l such that $m \mid l$ and $n \mid l$.
- Recall that we can use the prime factorization of numbers to calculate the GCD and the LCM:
- For example, $42 = 2 \cdot 3 \cdot 7$, $72 = 2^3 \cdot 3^2$, so $\gcd(42, 72) = 2 \cdot 3 = 6$ and $\text{lcm}(42, 72) = 2^3 \cdot 3^2 \cdot 7 = 504$.

7.1 Example Questions

Problem 7.1 For each of the following sets of numbers, find the GCD and LCM.

(a) 15 and 21.

Copyright © ARETEEM INSTITUTE. All rights reserved.

(b) 50, 90.

(c) 49, 154.

(d) 8, 20, 24.

Problem 7.2 Product of GCD and LCM

(a) What do you notice about the quantity $\gcd(m,n) \times \text{lcm}(m,n)$?

(b) Explain why the observation in part (a) is always true.

Problem 7.3 How many positive two-digit integers are factors of both 2240 and 2880? What are they?

Problem 7.4 At A.R.Teem Institute, there are 780 students in total, some of whom participate in Math Challenge classes. Among the Math Challenge class students, exactly $\frac{8}{17}$ are 6th graders, and exactly $\frac{9}{23}$ are 7th graders. How many students at A.R.Teem Institute do not attend Math Challenge classes?

Problem 7.5 Consider numbers that leave a remainder of 2 when divided by 3, 4, 5, and 6.

(a) Find the smallest such number.

7.1 Example Questions

(b) Find the largest such three-digit number.

Problem 7.6 In a Math Challenge class, the number of students is between 20 and 30. These students sit around a circular table, and start counting off numbers, clockwise, beginning with 1, and continue until 200 is counted. If the numbers 2 and 200 are counted by the same student, how many students are in the class?

Problem 7.7 In a math class, the teacher brings some pencils into the classroom. If he distributed the pencils evenly among the girls, each girl would get 15 pencils. If he distributed the pencils evenly among the boys, each boy would get 10 pencils. In fact, the teacher distributes the pencils evenly among all the students. How many pencils does each student receive?

Problem 7.8 Suppose A, B, C are integers ≥ 2 with (i) $\gcd(A, B) = 12$, (ii) $\text{lcm}(A, B) = 396$, and (iii) $\gcd(B, C) = 33$.

(a) What is $A \times B$?

(b) Calculate $\gcd(5A, 5B)$ and $\text{lcm}(5A, 5B)$.

(c) Calculate $\gcd(11A, B)$.

(d) What are the possibilities for $\gcd(A, C)$?

Problem 7.9 Suppose you live in a society where you only have $4 and $7 dollar bills.

(a) Show that it is impossible to pay for something (using only $4 and $7 dollar bills) that costs 17 dollars.

(b) Show that it is possible to pay for something (using only $4 and $7 dollar bills) that costs 18, 19, 20, or 21 dollars.

(c) Expand your argument from part (b) to show that $17 is the largest amount you cannot pay for using only $4 and $7 dollar bills.

Problem 7.10 There are 9 divisors for number A and 10 divisors for number B. The least common multiple of A and B is 2800. What are these two numbers?

7.2 Quick Response Questions

Problem 7.11 Use divisibility rules to help factor 4356. What is the largest prime number in the prime factorization of 4356?

Problem 7.12 Use divisibility rules and the Sieve of Eratosthenes to help factor 12345. What is the largest prime number in the prime factorization of 12345?

Problem 7.13 How many factors does 12348 have?

Problem 7.14 Find the prime factorization of 165 and 182. Find the sum of the largest primes in the prime factorizations.

Problem 7.15 Find the GCD of 165 and 182.

Problem 7.16 Find the LCM of 165 and 182.

Problem 7.17 Find all factors of 24. What is the sum of all factors of 24?

Problem 7.18 What is the LCM of 175 and 200?

Problem 7.19 Find the sum of all common multiples of 175 and 200 that divide 7000.

Problem 7.20 What is $\gcd(5 \cdot 42, 30) \div \gcd(42, 30)$?

7.3 Practice Questions

Problem 7.21 Find the GCD and LCM of $105, 231, 273$.

Problem 7.22 Suppose l, m, n are integers. Is it always true that $l \times m \times n = \gcd(l, m, n) \times \text{lcm}(l, m, n)$? Explain your answer and give at least one example.

Problem 7.23 Find all common factors of 504 and 540 that are perfect squares.

Problem 7.24 A math class with 20 students take an exam. A fourth of the students who passed got an A and a third who passed got a B. How many students passed the exam?

Problem 7.25 Find the largest three digit integer that has a remainder of 11 when divided by 15, 21, and 35.

Problem 7.26 A jar contains 99 jelly beans. Some students pass the jar around clockwise and eat a jelly bean every time they get the jar. If the 8th friend is the last one to get a jelly bean, how many friends are there?

Problem 7.27 Suppose an uncle distributes $1 bills among his nieces and nephews. If he distributed the dollars evenly among the nieces, each niece would get 24 dollars, and if he distributed them evenly among the nephews, each nephew would get 40 dollars. In fact, he distributes the dollars evenly among all his nieces and nephews. How many dollars does each niece or nephew receive?

Problem 7.28 Suppose A, B, C are integers ≥ 2 with (i) $\gcd(A,B) = 5$, (ii) $\text{lcm}(A,B) = 510$, and (iii) $\gcd(B,C) = 51$. What is $\gcd(17A, B)$?

Problem 7.29 What is the largest amount you cannot pay for using only $3 and $5 dollar bills?

Problem 7.30 Suppose that A has 9 divisors and B has 4 divisors. Find these numbers if $\gcd(A,B) = 7$ and $\text{lcm}(A,B) = 2205$.

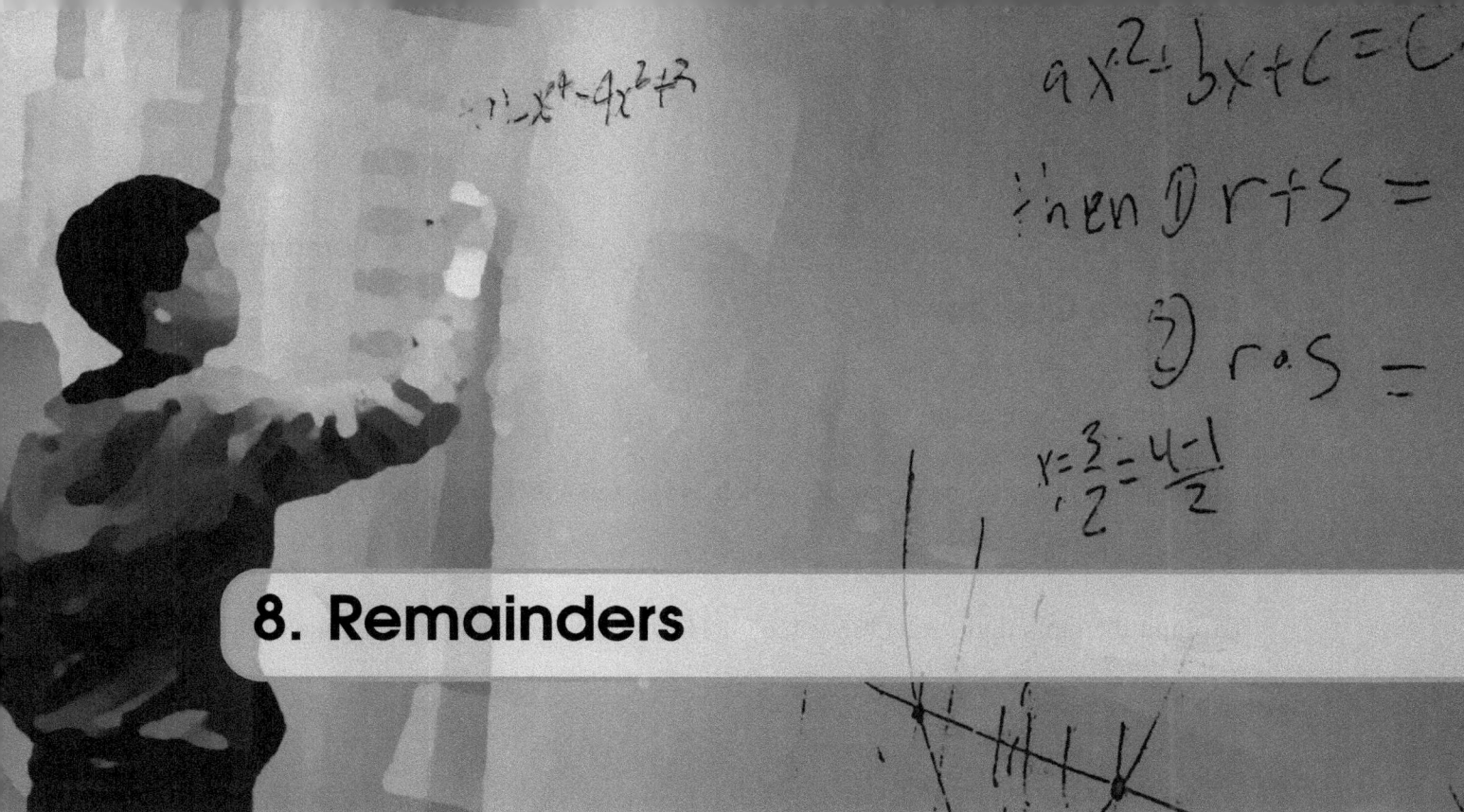

8. Remainders

Remainders

- For positive integers a and b, we can always compute $a \div b$, but the result (quotient) might not be an integer.
- We can, however, always write $a = bq + r$, where q is the quotient and r is the remainder, where $0 \leq r < b$.
- For example, $38 \div 5 = 7.6$. We can write $38 = 5 \cdot 7 + 3$, where 7 is the quotient and 3 is the remainder.

Modular Arithmetic

- The calculations and analyses about remainders is called *Modular Arithmetic*.
- If two numbers have the same remainder when divided by the same number m, we say they are *congruent modulo m*.
- For example, the numbers 9 and 2 are congruent modulo 7. Similarly, the numbers 17 and 3 are congruent modulo 7.
- In general, we can simply carry out the arithmetic operations using the remainders, and then if the results are greater than the modulus, we take remainders again. Then the remainders will be correct. We'll explore this in a little more detail next week.

Copyright © ARETEEM INSTITUTE. All rights reserved.

8.1 Example Questions

Problem 8.1 Units Digit

(a) Find the quotients and remainders when 34567 and 45678 are divided by 10?

(b) Find the units digit of $34567 + 45678$. How does it compare to the units digit of $7 + 8$.

(c) Find the units digit of 34567×45678. How does it compare to the units digit of 7×8.

(d) Explain your answers from part (b) and (c).

Problem 8.2 Everyday Problems with Remainders

(a) Suppose that the date is Saturday March 26th. What day of the week will March 26th be next year? (Assume next year is not a leap year.)

(b) Suppose it is 9 o'clock now. What time will it be 100 hours from now, if we ignore am/pm?

(c) What might be a better way to think of a $1000°$ angle?

Problem 8.3 Day of the Week Problems

8.1 Example Questions

(a) In a leap year, February is a month that contains Friday the 13th, what day of the week is March 1?

(b) Carlos Montado was born on Saturday, November 9, 2002. On what day of the week will Carlos be 706 days old?

Problem 8.4 Patterns!

(a) Find the units digit of 2^{2016}.

(b) Find the remainder when 2^{2016} is divided by 7.

Problem 8.5 When 1999^{2000} is divided by 5, what is the remainder?

Problem 8.6 What is the units digit of $19^{19} + 99^{99}$?

Problem 8.7 Each principal of Lincoln High School serves exactly one 3-year term. What is the maximum number of principals this school could have during an 8-year period?

Problem 8.8 The product of the two 9-digit numbers 404040404 and 707070707 has thousands digit A and units digit B. What is the sum $A + B$?

Problem 8.9 A group of pirates went to hunt for treasure. They found a chest of gold coins. They tried to equally divide the coins, but 6 coins were left over. So they picked one pirate among themselves and threw him overboard. Then they tried to divide the coins again, but 5 coins were left over. If the chest held 83 coins, how many pirates were there originally?

Problem 8.10 Let D be an integer greater than 1. When each of the three numbers 108, 201, and 356 is divided by D, the remainders are the same number R. Compute the value of $D - R$.

8.2 Quick Response Questions

Problem 8.11 What is the remainder when 120 is divided by 9?

Problem 8.12 What is the remainder when 492 is divided by 100?

Problem 8.13 What is the remainder when 787 is divided by 100?

Problem 8.14 Compare the remainders when $92 + 87$ and $492 + 787$ are divided by 100. Are the remainders equal?

Problem 8.15 What is the remainder when 100 is divided by 7?

Problem 8.16 What is the remainder when 492×787 is divided by 10?

Problem 8.17 What is the remainder when 5^{2017} is divided by 100?

Problem 8.18 What is the units digit of 2^{100}?

Problem 8.19 What is the remainder when 2^{100} is divided by 7?

Problem 8.20 What is the smallest number that leaves a remainder of 2 when you divide by 9 and by 11?

8.3 Practice Questions

Problem 8.21 Find the last two digits of 510137×486529.

Problem 8.22 Express 2000 hours in terms of weeks (w), days (d) and hours (h).

Problem 8.23 The vernal (spring) equinox occurs on March 20. This year March 20 was a Sunday. What day of the week will it be next year?

Problem 8.24 Find the units digit of 3^{2015}.

Problem 8.25 What is the remainder of 2014^{2015} when dividing by 9?

Problem 8.26 What is the units digit of $7^{47} + 7^{74}$?

Problem 8.27 Suppose Lincoln High School has an assistant principal that serves for exactly one 2-year term. What is the maximum number of principals and assistant principals this school could have during an 8-year period?

Problem 8.28 Find the sum of the last three digits of $123456789 \times 987654321$.

Chapter 8. Remainders

Problem 8.29 A group of pirates went to hunt for treasure. They found a chest of gold coins with 83 coins in it. They tried to equally divide the coins, but some coins were left over. So they picked one pirate among themselves and threw him overboard. Suppose the pirates continue in this manner (throwing each other overboard one by one) until they can divide the coins equally among themselves. How many pirates would be left?

Problem 8.30 Suppose that when the numbers 513, 571, 658 are divided by an integer $D > 1$, they all have the same remainder R. Find D and R.

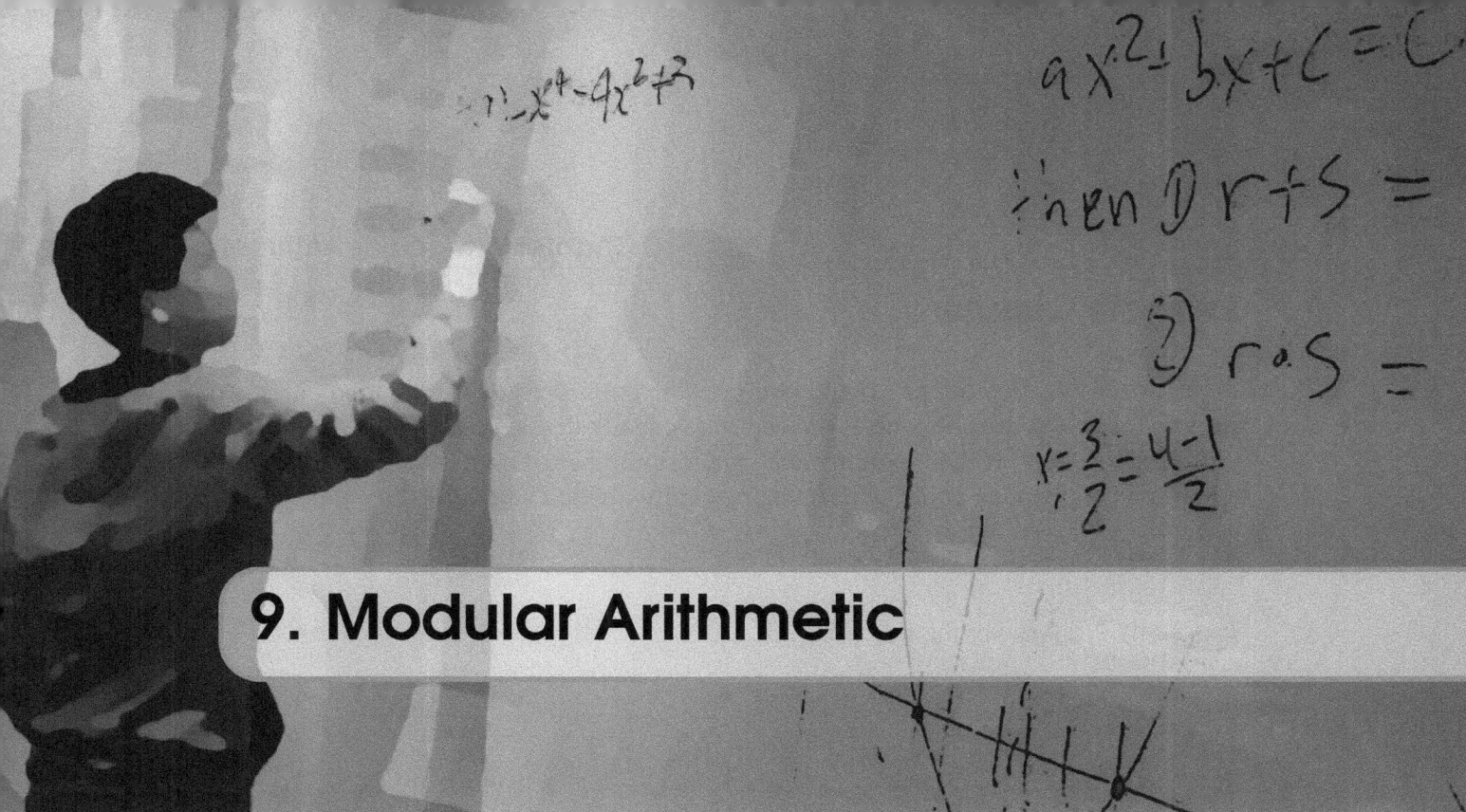

9. Modular Arithmetic

Modular Arithmetic Review

- If two numbers have the same remainder when divided by the same number m, we say they are *congruent modulo m*.
- In symbols, we write $a \equiv b \pmod{m}$ if a and b are congruent modulo m.
- Equivalently, if two numbers a, b are congruent modulo m ($a \equiv b \pmod{m}$), then $a - b$ must be divisible by m ($m \mid (a - b)$).
- Recall equality, addition, and multiplication all "work" with modular arithmetic. That is, we can simply carry out the arithmetic operations using the remainders, and then if the results are greater than the modulus, we take remainders again. Then the remainders will be correct.
- Modular arithmetic works the same way for negative integers as well.
- For example, $-7 \equiv 6 \pmod{13}$ and similarly $100 \equiv -4 \equiv 9 \pmod{13}$.

Divisibility Rules Revisited

- If a number is even, the remainder when dividing by 2 is 0. If a number is odd, the remainder when dividing by 2 is 1.
- When looking for the remainder when dividing by 10, it is simply the last/units digit.
- When looking for the remainder when dividing by 9, it is the same as the remainder of the sum of the digits divided by 9.
- When looking for the remainder when dividing by 3, it is the same as the remainder of the sum of the digits divided by 3.

Copyright © ARETEEM INSTITUTE. All rights reserved.

9.1 Example Questions

Problem 9.1 A box contains gold coins. If the coins are equally divided among six people, four coins are left over. If the coins are equally divided among five people, three coins are left over. If the box holds the smallest number of coins that meets these two conditions, how many coins are left when equally divided among seven people?

Problem 9.2 Answer the following.

(a) What is the units digit of $3^{215} + 7^{121}$?

(b) Find the remainder when $7^{18} + 9^{18}$ is divided by 8.

Problem 9.3 If $m > 1$ and $60 \equiv 70 \equiv 85 \pmod{m}$, what is m?

Problem 9.4 Find the last two digits of

(a) Find the last two digits of 99^{2016}.

(b) Find the last two digits of 7^{2015}.

Problem 9.5 Find the remainder when

(a) $35^3 + 53^3$ is divided by 10.

9.1 Example Questions

(b) $7^{18} + 9^{18}$ is divided by 8.

Problem 9.6 Modulo 9: Add the digits!

(a) Consider the number 1234. What is the quotient and remainder when 1234 is divided by 9?

(b) Consider the sum of the digits of 1234: $1+2+3+4 = 10$. What 10 (mod 9)?

(c) Consider a scrambled and combined sum of the digits of 1234: $34+21 = 55$. What is 55 (mod 9)?

(d) Summarize the results in this problem.

Problem 9.7 Suppose A, B, C, D are 4 consecutive natural numbers.

(a) Find the remainder when $A+B+C+D$ is divided by 4.

(b) Suppose you also know that $A+B+C+D$ is a three-digit number between 400 and 440 and $A+B+C+D$ is divisible by 9. Find A, B, C, D.

Problem 9.8 Concatenate the positive integers $1, 2, 3, \ldots, 2016$ to form a new integer:

$$12345678910111213 14 \cdots 201420152016.$$

What is the remainder when this new integer is divided by 9?

Problem 9.9 Three numbers, 22, 41, and 60 are divided by a positive integer d, and the three remainders are r_1, r_2, r_3 respectively. Given that $r_1 + r_2 + r_3 = 21$, determine the number d.

Problem 9.10 Initially, the positive integers $1, 2, 3, \ldots, 2016$ are written on the blackboard. Perform the following operation: At each step, three randomly chosen numbers on the board are erased, and replaced with the last digit of the sum of the three numbers. For example, if the numbers erased are 5, 27, and 2001, then write 3 on the board. Or, the numbers erased are 243, 62, and 334, then write 9 on the board. After many steps, only two numbers are left on the board. One is 48. What is the other number?

9.2 Quick Response Questions

Problem 9.11 Find the GCD of 50 and 60.

Problem 9.12 Find the LCM of 50 and 60.

Problem 9.13 What is the remainder when 100 is divided by 15?

Problem 9.14 What is the remainder when 88888888 is divided by 5?

Problem 9.15 What is the remainder when 88888888 is divided by 9?

Problem 9.16 What is the units digit of 2^{100}?

Problem 9.17 Find the last two digits of 216×348.

Problem 9.18 Is it true that $178 \equiv 48 \pmod{11}$?

Problem 9.19 Is it true that $178 \equiv 48 \pmod{13}$?

Problem 9.20 Is it true that $1234567 \equiv 7654321 \pmod{3}$?

9.3 Practice Questions

Problem 9.21 A box contains gold coins. If the coins are equally divided among seven people, five coins are left over, if they are equally divided among five people, three coins are left over, and if they are evenly divided among three people, one coin is left over. If the box holds the smallest number of coins that meets these two conditions, how many coins are left when equally divided among nine people?

Problem 9.22 Find the remainder of $2^{102} + 13^{102}$ when it is divided by 15.

Problem 9.23 If $m > 1$ and $17 \equiv 56 \equiv 108 \pmod{m}$, what is m?

Problem 9.24 Find the last two digits of 21^{2014}.

Problem 9.25 Find the remainder when $3^{15} - 7^{15}$ is divided by 5.

Problem 9.26 What is the remainder when

$$1 + 22 + 333 + 4444 + 55555$$

is divided by 9?

Problem 9.27 Find the set of four smallest positive consecutive integers A, B, C, D such that $A + B + C + D$ is divisible by 9 and 11.

Chapter 9. Modular Arithmetic

Problem 9.28 Is the integer $1 + 2 + 3 + \cdots + 2014 + 2015 + 2016$ divisible by 11?

Problem 9.29 Find all $m > 1$ such that $1023 \equiv 561 \pmod{m}$.

Problem 9.30 Initially the positive integers $1, 2, 3, \ldots, 2017$ are written on the blackboard. Perform the following operation: At each step, two randomly chosen numbers are erased, and replaced with the last digit of their sum. For example, if the numbers erased are 58 and 197, then write 5 on the board. Or, if the numbers 294 and 710 are erased, then write 4 on the board. After many steps, only one number is left on the board. What is this number?

Copyright © ARETEEM INSTITUTE. All rights reserved.

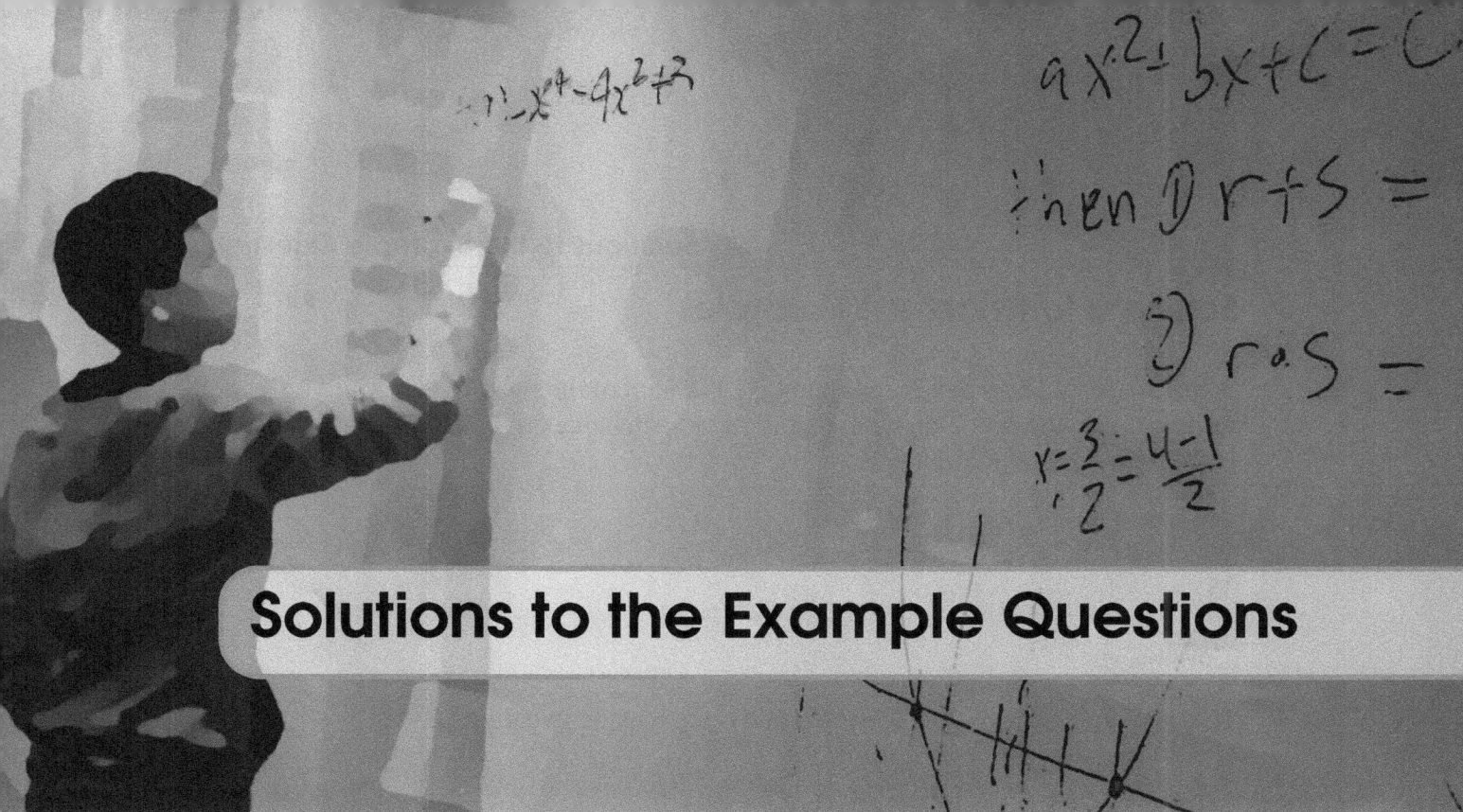

Solutions to the Example Questions

In the sections below you will find solutions to all of the Example Questions contained in this book.

Quick Response and Practice questions are meant to be used for homework, so their answers and solutions are not included. Teachers or math coaches may contact Areteem at info@areteem.org for answer keys and options for purchasing a Teachers' Edition of the course.

1 Solutions to Chapter 1 Examples

Problem 1.1 There is a 2-digit number. The sum of the two digits is 8. If you switch the two digits, the new 2-digit number is 18 more than the original 2-digit number. What is the original 2-digit number?

Answer

35

Solution

There are several ways that we could add up to 8 using two digits:

$$0 \& 8, \quad 1 \& 7, \quad 2 \& 6, \quad 3 \& 5, \quad \text{and} \quad 4 \& 4.$$

Note that we don't want to use 0, because if we switch the digits of a number that ends in 0, we get a one-digit number. So, our number should be one of 17, 71, 26, 62, 35, 53 or 44. We know that when we switch the numbers we get a bigger number than the original, that means that it should be either 17, 26 or 35 since the other ones become smaller (or stay the same) when we swap the digits. If we look at the differences between those numbers and their "switched" versions, we get: $71 - 17 = 57$, $62 - 26 = 36$ and $53 - 35 = 18$. So the number we are looking for is 35.

Problem 1.2 Find a 6-digit number that has no repeated digits and that multiplied by its last digit is equal to 999999.

Answer

142857

Solution

The ones' digit of our number has to be either 3 or 7, since those are the only 1-digit numbers that end in 9 when we square them. So the 6-digit number we are looking for will be either $999999 \div 3 = 333333$ or $999999 \div 7 = 142857$. Since we cannot repeat digits, the number we want is 142857.

Problem 1.3 Do you get an odd number or even number if you add 9 even numbers?

1 Solutions to Chapter 1 Examples

Answer

Even

Solution

Remember, all even numbers end up in either 0, 2, 4, 6 or 8. Every time we add up two numbers that end with those digits, we will get again a number that ends up in either 0, 2, 4, 6 or 8. That means *every time* we add up to even numbers we get again an even number, so, if we do that 9 times we will still get an even number.

Problem 1.4 Uncle Jim got lost while we were driving back to California from Montréal. He saw a sign that said how many miles away we were from Los Angeles. Since he was driving fast, he couldn't quite see the number, but he knew it had 4 digits. I saw the number, but I wanted to have some fun so instead of telling him the number right away I gave him some clues:

- The number has the digit 1 somewhere.
- The digit in the hundreds' place is three times the digit in the thousands' place.
- The digit in the ones' place is 4 times the digit in the tens' place.
- The thousands' digit is 2.

How far away are we from Los Angeles?

Answer

2614 miles

Solution

The first digit will be 2 and the second digit will be $2 \times 3 = 6$. Since we haven't used a 1 yet, we must have 1 ten and $1 \times 4 = 4$ ones. Hence the number in the sign is 2614.

Problem 1.5 Do the following calculations:

(a) $92 + 86 + 91 + 88 + 87 + 90 + 89 + 93 + 92 + 88$.

Answer

896

Solution

We can rewrite the sum using numbers that would give us the original numbers if we add them to, or subtract them from, 90. In this case, our quick sum would look like:

$$2 - 4 + 1 - 2 - 3 + 0 - 1 + 3 + 2 - 2 = -4.$$

Since we had 10 numbers, we will need to subtract that 4 from $10 \times 90 = 900$. So, we have that the total sum is $900 - 4 = 896$.

(b) $54 + 67 + 33 + 84 + 46 + 64 + 16$.

Answer

364

Solution

Notice that some numbers add up to 100. We can spot them easily because their units' digits add up to 10 and their tens' digits add up to 9.

$$\underline{54} + \underline{67} + \underline{33} + \underline{84} + \underline{46} + 64 + \underline{16}$$

This means we can add the numbers a lot quicker if we replace each of those pairs with a 100:

$$100 + 100 + 100 + 64 = 364.$$

Problem 1.6 In the following calculations try to simplify before multiplying and dividing.

(a) $213 \times 24 \div 12$

Answer

426

Solution

If we were to do this operations in order from left to right we would work with big numbers:

$$213 \times 24 \div 12 = 5112 \div 12 = 426.$$

Copyright © ARETEEM INSTITUTE. All rights reserved.

1 Solutions to Chapter 1 Examples

We want to try to spot something we can do to work with smaller numbers. Note that we have a 24 that is multiplying and a 12 that is dividing. Since $24 \div 12 = 2$, if we do that first we will end up with

$$213 \times 24 \div 12 = 213 \times 2 = 426.$$

(b) $75 \times 32 \div 400$

Answer

6

Solution

$$\frac{75 \times \cancel{32}^{16}}{\cancel{400}_{200}} = \frac{75 \times \cancel{16}^{8}}{\cancel{200}_{100}} = \frac{75 \times \cancel{8}^{4}}{\cancel{100}_{50}} = \frac{75 \times \cancel{4}^{2}}{\cancel{50}_{25}} = \frac{\cancel{75}^{3} \times 2}{\cancel{25}} = 3 \times 2 = 6$$

(c) $36 \times 62 \div 2 \times 7 \div 3$

Answer

2604

Solution

This time we have more than one number dividing. Let's put the numbers in two groups: in one group all the numbers that are multiplying and on the other all the numbers that are dividing. For convenience we will write the two groups of numbers above and below a horizontal line, that way we can spot easier if there is something we can do first.

$$\frac{36 \times 62 \times 7}{2 \times 3}.$$

We can divide 36 by 3 and get 12 instead, and we can divide 62 by 2 and get 31 instead. So we have

$$\frac{\cancel{36}^{12} \times 62 \times 7}{2 \times \cancel{3}} = \frac{12 \times \cancel{62}^{31} \times 7}{\cancel{2}} = 12 \times 31 \times 7 = 372 \times 7 = 2604$$

Copyright © ARETEEM INSTITUTE. All rights reserved.

We still had to deal with some big numbers in here, however, the numbers we worked with in the end were a lot smaller than if we had performed the operations in the order they were originally:

$$36 \times 62 \div 2 \times 7 \div 3 = 2232 \div 2 \times 7 \div 3 = 1116 \times 7 \div 3 = 7812 \div 3 = 2604$$

Problem 1.7 What do we get if we multiply each of the following numbers by 11?

(a) 253

Answer

2783

Solution

When we multiply a number by 11, the last digit is the last digit of the original number, and the first digit is the first digit of the original number. The rest of the digits can be obtained by adding neighbors together. So in 253×11 the first digit will be 2, the last digit will be 3, and the digits in between will be $2 + 5 = 7$ and $5 + 3 = 8$.

$$\begin{array}{ccccc} & 2 & 5 & 3 & \\ & \swarrow \searrow & \swarrow \searrow & \swarrow \searrow & \\ 2 & 2+5 & 5+3 & & 3 \\ 2 & 7 & 8 & & 3 \end{array}$$

So we get
$$253 \times 11 = 2783.$$

(b) 3594

Answer

39534

Solution

Sometimes, when we add up neighbors we will get a number that is bigger than 9, so it

Copyright © ARETEEM INSTITUTE. All rights reserved.

1 Solutions to Chapter 1 Examples

has more than one digit. In those cases we will need to carry over the extra digit to the *left* neighbor sum.

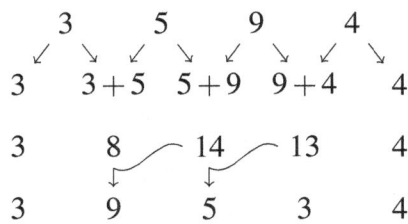

(c) 45729

Answer

503019

Solution

If after carrying over the extra 1s we get once more more than 9 for one of the digits, we simply carry it over again.

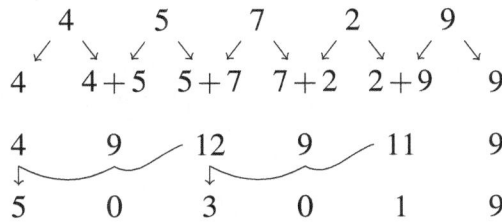

Problem 1.8 Square numbers that end in 5

(a) 15^2

Answer

225

Solution

When squaring a number that ends in 5, the number we get will *always* end in 25 because

$5 \times 5 = 25$. We can figure the rest of the digits by multiplying the number we get by removing the 5 together with its successor (that is, the next counting number).

In the number 15 when we remove the 5, we are left with 1. So we need to multiply $1 \times 2 = 2$ and then attach 25 at the end of it:

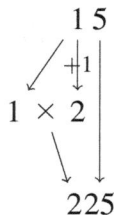

This means $15^2 = 225$.

(b) 165^2

Answer

27225

Solution

The number 165 without the 5 at the end is just 16, and the successor of 16 is 17. So, we need to multiply $16 \times 17 = 272$ and attach a 25 at the end:

$$\begin{array}{c} 165 \\ \diagup \ \ \downarrow{+1} \\ 16 \times 17 \\ \diagdown \ \ \downarrow \\ 27225 \end{array}$$

(c) 2005^2

Answer

4020025

Copyright © ARETEEM INSTITUTE. All rights reserved.

1 Solutions to Chapter 1 Examples

Solution

This strategy also works with big numbers. If we remove the 5 at the end of 2205 we are left with 220, so to find 2005^2 we would just need to multiply $200 \times 201 = 40200$ and attach a 25 at the end of it:

$$2005$$
$$200 \times 201$$
$$4020025$$

So, $2005^2 = 4020025$.

Problem 1.9 Multiply numbers that start with the same digits and end in digits that add up to 10.

(a) 48×42

Answer

2016

Solution

When the numbers we are multiplying start the same way and their last digits add up to 10, something similar to when we square numbers that end up in 5 will happen. The last two digits of the product will be the product of the last digits of the numbers (similar to the last digits being $5 \times 5 = 25$). The first digits of the product will be given by multiplying the starting digits by their successor. So, 48×42 will be given by

$$\begin{aligned} &= 4 \times 5 \quad 8 \times 2 \\ &= \quad 20 \quad\quad 16 \end{aligned}$$

(b) 39×31

Answer

1209

Solution

When the last digits are 1 and 9 we want to be careful. The answer will not end just in 9, but in 09. This is because the product of the first digit and their successor represents the hundreds of our number, so we will have $3 \times 4 = 12$ hundreds and 9 ones, so 1209.

(c) 264×266

Answer

70224

Solution

$26 \times 27 = 702 \quad 4 \times 6 = 24$

Problem 1.10 Troy bought some candies at the local grocery store in town. He had to pay $6.23, so he gave the cashier a $10 bill. The cashier gave him back three $1 bills and... 77 pennies! It seems there were no other coins available in the cash register. No one likes pennies and neither does Troy. He looked in his pockets for some coins that he could use instead, and he found 1 dime, 3 quarters, 2 pennies and 2 nickels. What coins will help him get the smallest amount of pennies back?

Answer

1 quarter

Solution

If Troy used the dime, the two nickels and the two pennies he would get

$$10 + 2 \times 5 + 2 \times 1 = 22$$

cents in total, but he needs at least 23¢ to avoid getting all those pennies back. By just using one quarter he gets 25 ¢, which is just a little bit more than the 23¢ he needs, so if he gives the cashier the $10 bill and one quarter, the cashier would have to give him back

$$10.25 - 6.23 = 4.02$$

dollars, that is, four $1 bills and two pennies.

Copyright © ARETEEM INSTITUTE. All rights reserved.

2 Solutions to Chapter 2 Examples

Problem 2.1 (Checking Whether a Number is Prime)

(a) Is the number 37 prime? How can you prove this?

Answer

Yes

Solution

Check that 37 is not divisible by $2, 3, 4, \ldots, 36$. In fact, we only need to check $2, 3, 5$, why?

(b) Expand your method in part (a) to work for any integer N. Explain why it works!

Solution

In general, to check if a number N is prime, it is enough to check if N is divisible by any integer (or really prime numbers) between 2 and $\sqrt{N}$.

Problem 2.2 (Sieve of Eratosthenes) One of the earliest methods for finding prime numbers is called the *sieve of Eratosthenes*. It works like this: Start with a table of numbers; Cross out the number 1; Keep the number 2 as a prime, and cross out all other multiples of 2 (the even numbers); Keep the next number, 3, as a prime, and cross out all other multiples of 3; Keep the next number, 5, as a prime, and cross out all other multiples of 5; and so on. At the end, all remaining numbers on the table are primes.

(a) Using the table below, find all the prime numbers less than 180.

1	2	3	4	5	6	7	8	9	10	11	12
13	14	15	16	17	18	19	20	21	22	23	24
25	26	27	28	29	30	31	32	33	34	35	36
37	38	39	40	41	42	43	44	45	46	47	48
49	50	51	52	53	54	55	56	57	58	59	60
61	62	63	64	65	66	67	68	69	70	71	72
73	74	75	76	77	78	79	80	81	82	83	84
85	86	87	88	89	90	91	92	93	94	95	96
97	98	99	100	101	102	103	104	105	106	107	108
109	110	111	112	113	114	115	116	117	118	119	120
121	122	123	124	125	126	127	128	129	130	131	132
133	134	135	136	137	138	139	140	141	142	143	144
145	146	147	148	149	150	151	152	153	154	155	156
157	158	159	160	161	162	163	164	165	166	167	168
169	170	171	172	173	174	175	176	177	178	179	180

Solution

The primes less than 180 are 2, 3, 5, 7, 11, 13, 17, 19, 23, 29, 31, 37, 41, 43, 47, 53, 59, 61, 67, 71, 73, 79, 83, 89, 97, 101, 103, 107, 109, 113, 127, 131, 137, 139, 149, 151, 157, 163, 167, 173, 179.

(b) Explain why the sieve of Eratosthenes works.

2 Solutions to Chapter 2 Examples

Solution

One by one, we cross of all multiples of 2 (except 2), all multiples of 3 (except 3), etc, so we are left with only the primes.

Problem 2.3 A "prime-prime" number is a prime number that yields a prime number when its units digit is omitted. For example, 131 is a three-digit "prime-prime" number because 131 is prime and 13 is prime.

(a) How many two-digit "prime-primes" are there?

Answer

9

Solution

They are $23, 29, 31, 37, 53, 59, 71, 73, 79$.

(b) Without actually counting them, argue that there "should" be more "prime-prime" numbers between $100 - 200$ than between $200 - 300$.

Solution

Note the prime numbers between $10 - 20$ are $11, 13, 17, 19$, but the only primes between $20 - 30$ are $23, 29$. Hence there are many more possibilities in the range $100 - 200$.

Problem 2.4 For each of the following numbers, determine whether the number is prime or not. If the number is not prime, give an example of a non-trivial divisor.

(a) 457

Solution

Prime.

(b) 301

Answer

Not prime.

Copyright © ARETEEM INSTITUTE. All rights reserved.

Solution

7, 43 are a divisors of 301.

(c) 657

Answer

Not prime.

Solution

3 is a divisor of 657.

(d) 247

Answer

Not prime.

Solution

13, 19 are divisors of 247.

Problem 2.5 Determine a value d such that $1d3$ and $5d7$ are prime numbers.

Answer

7

Solution

The numbers 173 and 577 are prime numbers. Using the primes we found in Problem 2.2 (if we want $1d3$ to be prime) we need to check $d = 0, 1, 6, 7, 9$. Remember that $510, 540, 570, 600$ are multiples of 3 so this removes $d = 0, 6, 9$. Then note $517 = 11 \cdot 47$. By process of elimination, d must be 7. (Double check that 577 is prime yourself!)

Problem 2.6 Consider the number 2016.

(a) Find the prime factorization of 2016.

2 Solutions to Chapter 2 Examples

Answer

$2^5 \cdot 3^2 \cdot 7$

Solution

$2016 = 2^5 \cdot 3^2 \cdot 7$.

(b) What is the sum of the *distinct* prime factors of 2016?

Answer

12

Solution

The distinct prime factors are $2, 3, 7$.

Problem 2.7 The numerical values of the years are favorite numbers of many math contests. The prime factorizations of these numbers are often useful in those contests. We already found the prime factorizations of 2016 in Problem 2.6 Find the prime factorizations of the following:

(a) 2017

Solution

2017 is prime.

(b) 2018

Solution

$2018 = 2 \cdot 1009$ (1009 is prime).

(c) 2019

Solution

$2019 = 3 \cdot 673$ (673 is prime).

(d) 2015

Solution

$2015 = 5 \cdot 13 \cdot 31$.

Problem 2.8 Let N be the largest seven-digit number that can be constructed using each of the digits $1,2,3,4,5,6,7$ such that the sum of every pair of consecutive digits is a prime number. What is the number N?

Answer

7652341

Solution

The largest number is 7654321. $7+6 = 13, 6+5 = 11$ (both prime) but $5+4 = 9$ which is not prime. The next largest is $5+2 = 7$ which is prime. Continue in this manner to get the full number.

Problem 2.9 Consider the numbers $17, 19, 21, 25, 26$.

(a) Which of the numbers has the smallest prime factor?

Answer

26

Solution

All the other numbers are odd, but 26 is even (so 2 is a factor).

(b) Which of the numbers has the largest prime factor?

Answer

19

Solution

Recall 19 is prime (none of the other numbers are prime), so 19 is the largest prime factor of any of the numbers.

Problem 2.10 Find the prime factorization of 510510.

Copyright © ARETEEM INSTITUTE. All rights reserved.

Answer

$510510 = 2 \cdot 3 \cdot 5 \cdot 7 \cdot 11 \cdot 13 \cdot 17$

Solution

Note 510510 is the product of the first 7 primes.

3 Solutions to Chapter 3 Examples

Problem 3.1 Even Number of Factors?

(a) Use the "pairing" trick mentioned above to help find and count all the factors of 72.

Answer

There are 12 factors: $1, 2, 3, 4, 6, 8, 9, 12, 18, 24, 36, 72$

Solution

Note that the pairs are $1, 72, 2, 36, 3, 24, 4, 18, 6, 12, 8, 9$, where each pair has product 72.

(b) We might be tempted after part (a) to say that every number has an even number of factors. (Or maybe tempted to say every number greater than 1 has an even number of factors.) Disprove this hypothesis by using the pairing trick to help find and count all factors of 36.

Answer

There are 9 factors: $1, 2, 3, 4, 6, 9, 12, 18, 36$

Solution

Note the pairs are 1, 36; 2, 18; 3, 12; 4, 9; and 6, 6; where each pair has product 36. Note that 6 is paired with itself, so there are an odd number of factors.

(c) Come up with a hypothesis for when a number n has an even number of factors. You do not need to prove this now.

Solution

Note that $6^2 = 36$, so we guess that if n is a perfect square it has an odd number of factors, otherwise it has an even number of factors.

Problem 3.2 Of the first 50 positive integers, which have an odd number of factors and which have an even number of factors?

Copyright © ARETEEM INSTITUTE. All rights reserved.

3 Solutions to Chapter 3 Examples

Answer

$\{1, 4, 9, 16, 25, 36, 49\}$ have an odd number, the rest have an even number

Solution

As we hypothesized above, only the perfect squares have an odd number of factors.

Problem 3.3 For any positive integer n, define $\boxed{n}$ to be the sum of the positive factors of n. For example, $\boxed{6} = 1 + 2 + 3 + 6 = 12$.

(a) Find $\boxed{20}$.

Answer

42

Solution

$\boxed{20} = 1 + 2 + 4 + 5 + 10 + 20 = 42$.

(b) Find $\boxed{\boxed{11}}$.

Answer

28

Solution

11 is prime, so $\boxed{11} = 1 + 11 = 12$. Then $\boxed{12} = 1 + 2 + 3 + 4 + 6 + 12 = 28$.

Problem 3.4 Counting Multiples Practice

(a) How many three-digit numbers are divisible by 17?

Answer

53

Solution

We have $999 \div 17 = 58 \; R \; 13$, so there are 58 numbers less than 1000 divisible by 17.

Similarly, $99 \div 17 = 5 \ R \ 17$, so 5 of them are less than 100. Hence $58 - 5 = 53$ are three-digit numbers.

(b) How many whole numbers strictly between 1 and 2009 are divisible by 2 but not by 7?

Answer

861

Solution

$2009 \div 2 = 1004.5$ so there are 1004 such numbers divisible by 2. To be divisible by 2 and 7, a number must be divisible by 14. There are 143 numbers ($2009 \div 14 = 143.5$) that are also divisible by 7. Hence in total there are $1004 - 143 = 861$ numbers.

Problem 3.5 Which number has more positive divisors, 2015 or 2020?

Answer

2020

Solution

$2015 = 5 \times 13 \times 31$ so it has $(1+1)(1+1)(1+1) = 8$ divisors. $2020 = 2^2 \times 5 \times 101$ so it has $(2+1)(1+1)(1+1) = 12$ divisors.

Problem 3.6 Find the smallest positive integer x such that

(a) $252 \cdot x$ is a perfect square.

Answer

7

Solution

Note $252 = 2^2 \cdot 3^2 \cdot 7$. To make a perfect square we only need another 7.

(b) $252 \cdot x$ is a perfect cube.

Copyright © ARETEEM INSTITUTE. All rights reserved.

3 Solutions to Chapter 3 Examples

Answer

294

Solution

As above $252 = 2^2 \cdot 3^2 \cdot 7$. To make a perfect cube we need $2 \cdot 3 \cdot 7^2 = 294$.

Problem 3.7 A prison houses 1,000 inmates in 1,000 prison cells. It also has 1,000 guards. One day the guards decide to use a number theory problem to free some prisoners. The first guard goes through the prison and unlocks every cell. The second guard goes through the prison and locks every second cell. The third guard goes through the prison and changes the status of every third cell. That is, if it's locked, he unlocks it; and if it's unlocked, he locks it. This process continues (that is the nth prison guard changes the status of every nth cell) until the 1,000th guard has passed through the prison, after which every prisoner whose cell is left unlocked is free to leave. How many prisoners will be set free?

Answer

31

Solution

Note that cell n will be left open if n has an odd number of factors. Hence, only cells whose number is a square will be unlocked. $31^2 = 961$ but $32^2 = 1024$, so there are 31 perfect squares less than 1000.

Problem 3.8 How many perfect squares are factors of 2592?

Answer

9

Solution

Note $2592 = 2^5 \cdot 3^4$. To form a factor that is a perfect square, we use even powers of $2, 3 : 2^0, 2^2, 2^4, 3^0, 3^2, 3^4$. There are $3 \cdot 3 = 9$ combinations of these which give the perfect squares as factors.

Problem 3.9 Find the smallest prime p such that $2^p - 1$ is not a prime number.

Answer

11

Solution

$2^2-1 = 3, 2^3-1 = 7, 2^5-1 = 31, 2^7-1 = 127$ are all prime numbers, but $2^{11}-1 = 2047 = 23 \times 89$ is composite.

Problem 3.10 From the set of natural numbers $2, 3, 4, \ldots, 999$, delete 9 subsets as follows. First delete all even numbers except 2, then all multiples of 3, except 3, then all multiples of 5, except 5, and so on for the nine primes 2, 3, 5, 7, 11, 13, 17, 19, and 23. After all these steps, there are only three composite numbers left. Find the sum of these composite numbers.

Answer

2701

Solution

Recall every number has a prime factorization. Since the smallest primes larger than 23 are $29, 31, 33, 37$. Hence the smallest possible composite numbers not deleted are $29^2, 29 \cdot 31, 29 \cdot 33, 31 \cdot 31, 31 \cdot 31$, etc. The only ones smaller than 1000 are $29^2 = 841, 29 \cdot 31 = 899, 31 \cdot 31 = 961$ with sum $841 + 899 + 961 = 2701$.

4 Solutions to Chapter 4 Examples

Problem 4.1 Come up with a divisibility rule for the following numbers. (That is, find an easy way to check if a number is divisible by the following numbers.) Can you explain why the rules work?

(a) 2.

Answer

Look at the last (units) digit. If the last digit is even, then the number is divisible by 2.

Solution

This works because 10 is always divisible by 2.

(b) 5.

Answer

Look at the last (units) digit. If the last digit is 0 or 5, then the number is divisible by 5.

Solution

This works because 10 is always divisible by 5.

Problem 4.2 Divisibility Rule for Three

(a) Write out the representation of the number 15624 in terms of its digits and place values.

Solution

$15624 = 1 \times 10000 + 5 \times 1000 + 6 \times 100 + 2 \times 10 + 4.$

(b) Show that "Is 15624 divisible by 3?" has the same answer as "Is $1+5+6+2+4$ divisible by 3?". Hint: What is 9×1? 9×11? 9×111? 9×1111?

Solution

We have

$$15624 = 1 \times 10000 + 5 \times 1000 + 6 \times 100 + 2 \times 10 + 4$$
$$= 1 \times (9999+1) + 5 \times (999+1) + 6 \times (99+1) + 2 \times (9+1) + 4$$
$$= 1 \times 9999 + 1 + 5 \times 999 + 5 + 6 \times 99 + 6 + 2 \times 9 + 2 + 4$$
$$= (1 \times 9999 + 5 \times 999 + 6 \times 99 + 2 \times 9) + (1 + 5 + 6 + 2 + 4).$$

Notice that the quantity in the first parentheses is divisible by 3 (in fact it is a multiple of 9). Thus whether or not 15624 is divisible by 3 has the same answer as whether or not $1+5+6+2+4$ is divisible by 3. In this case, both answers are "yes".

(c) Use part (b) to state a general divisibility rule for 3.

Solution

Add up all the digits. If the result is divisible by 3, then the original number is divisible by 3.

Problem 4.3 Come up with divisibility rules for the following numbers.

(a) 6

Answer

If the number is divisible by 2 and by 3, then it is divisible by 6.

Solution

Note that $2 \times 3 = 6$ and 6 is the smallest multiple of both 2 and 3.

(b) 4

Answer

Look at the last two digits. If the last two digits form a number divisible by 4, then the number is divisible by 4.

Solution

This works because 100 is always divisible by 4.

4 Solutions to Chapter 4 Examples

Problem 4.4 What five-digit multiple of 11 consists entirely of 2s and 3s?

Answer

23232

Solution

View the alternating sum of a five-digit number as two differences plus the final number. For example, the alternating sum of 32232 is $3-2+2-3+2 = (3-2)+(2-3)+2 = 1+(-1)+2 = 2$. The final digit (the units digit) is either 2 or 3. The two differences are each either $0, +1, -1$. Hence the only way to get a multiple of 11 is $(-1)+(-1)+2 = 0$. Therefore the five-digit multiple of 11 is 23232 (with alternating sum $2-3+2-3+2 = 0$).

Problem 4.5 A number $\overline{35a2a}$ is divisible by 3. Its last three digits form a three-digit number $\overline{a2a}$ that can be exactly divided by 2. Find this five-digit number.

Answer

35424

Solution

The digit a has to be even. The sum of digits $3+5+a+2+a = 10+2a$ is divisible by 3. The only digit a that meets this requirement is 4.

Problem 4.6 A four digit number $\overline{7a4b}$ is divisible by 18. Find the value of a and b so that this four-digit number has the least value.

Answer

$a = 1, b = 6$

Solution

The number should be divisible by both 2 and 9. We want b to be even and the sum $7+a+4+b$ to be a multiple of 9, and want to find the smallest such numbers. There are only a few possible choices, and we can work it out by trial and error, starting with small a. First try $a = 0$; in this case b must be 7 in order for the number to be divisible by 9, but 7 is not even. So we try $a = 1$, and get $b = 6$ and 7146 works.

Problem 4.7 Find all possible six-digit numbers represented by 2016__ __ which can exactly be divided by 2, 3, 4, 5.

Answer

201600, 201660

Solution

The last digit should be multiple of 5, and also even, so it has to be 0.

The number is divisible by 3, so the sum of the digits is a multiple of 3. The sum of digits $2+0+1+6 = 9$, Thus, the sum of the last two digits should be 0, 3, 6 or 9. Hence our possible numbers are 201600, 201630, 2016, 60, 201690. Only 201600, 201660 are divisible by 4 (as $4 \mid 0, 60$ and $4 \nmid 30, 90$).

Problem 4.8 For how many positive integer values of n are both $\dfrac{n}{3}$ and $3n$ three-digit integers?

Answer

12

Solution

$3n \leq 999$, so $n \leq 333$. Also, $n/3 \geq 100$, so $n \geq 300$. We need multiples of 3 between 300 and 333, inclusive. There are 12 of them.

Problem 4.9 How many three-digit numbers are divisible by 13?

Answer

69

Solution

The following is the most accurate method: $999/13 \approx 76.8$, thus from 1 to 999, there are $\lfloor 999/13 \rfloor = 76$ multiples of 13. $99/13 \approx 7.6$, thus from 1 to 99 there are 7 multiples of 13. Thus the number of 3-digit multiples of 13 is $76 - 7 = 69$.

4 Solutions to Chapter 4 Examples

Problem 4.10 A natural number is a multiple of both 3 and 4, and has totally 10 divisors. What is this number?

Answer

48

Solution

Based on the formula for the number of divisors, we first find out the factorization of 10: $10 = 2 \times 5 = (1+1)(4+1)$. So the number in question should have the form $p \times q^4$, or p^9, where p and q represent prime numbers. But we already know that this number is a multiple of 3 and 4, so it has prime factors 2 and 3, and cannot be of the form p^9. Thus it should be of the form $p \times q^4$. Since $4 = 2^2$, p cannot be 2, so the only possibility is $3 \times 2^4 = 48$.

5 Solutions to Chapter 5 Examples

Problem 5.1 What digit could fill in the blank to make 89__43

(a) divisible by 11?

Answer

2

Solution

Divisibility by 11 is verified by using the alternating sum of the digits. We have $8-9+?-4+3 = ?-2$. Since we want the alternating sum to be a multiple of 11, the only possible missing digit is 2.

(b) divisible by 9?

Answer

3

Solution

Divisibility by 9 is verified by using the sum of the digits. We have $8+9+?+4+3 = 24+?$. Since we want the sum of be a multiple of 9, the only possible missing digit is 3.

(c) divisible by 4?

Answer

None

Solution

Divisibility by 4 is verified by using the last two digits. However, 43 is not divisible by 4, so no matter what digit we use in the blank, the number will not be divisible by 4.

Problem 5.2 What is the smallest positive integer,

Copyright © ARETEEM INSTITUTE. All rights reserved.

5 Solutions to Chapter 5 Examples

(a) that is divisible by 2 and 3 that consists entirely of 2's and 3's, with at least one of each?

Answer

2232

Solution

It is divisible by 2, so the last digit must be 2. It is divisible by 3, so the sum of the digits is a multiple of 3, thus there should be at least three 2s. The digit 3 has to appear once, and the smallest possible position for this 3 is the tens place.

(b) entirely consisting of 4 and 9, at least one of each, that is divisible by both 4 and 9?

Answer

4444444944

Solution

To be divisible by 4, the last two digits need to be divisible by 4. The smallest such last 2 digits are thus 44. To be divisible by 9, the sum of the digits must be divisible by 9, so (since 9 is already divisible by 9) there must be at least nine 4's. The digit 9 has to appear once, and the smallest possible position for this is 9 in the hundreds place.

Problem 5.3 How many positive multiples of 3 less than 1000 use only the digits 2 and/or 4?

Answer

4

Solution

Note the sum of the digits must be a multiple of 3. The small multiples of 3 that can be written as a sum of 2's and 4's are $6 = 2+4 = 2+2+2$ and $12 = 4+4+4$. Then they can be listed: 24, 42, 222, 444.

Problem 5.4 A four-digit number $\overline{7x4y}$ should be exactly divisible by 55. Find all possible four-digit numbers that meet this requirement.

Answer

7040, 7645

Solution

Since $55 = 5 \times 11$, we want the number to be divisible by 5 and 11. Hence the last digit must be either 0 or 5. If the last digit is 0, then we need $7 - x + 4 - 0 = 11 - x$ to be a multiple by 11. In this case $x = 0$ is the only possibility. If the last digit is 5, then we need $7 - x + 4 - 5 = 6 - x$. In this case $x = 6$ is the only possibility.

Problem 5.5 The number 64 has the property that it is divisible by its units digit. How many whole numbers between 10 and 50 have this property?

Answer

17

Solution

They can be listed based on their last digits: 11, 21, 31, 41; 12, 22, 32, 42; 33; 24, 44; 15, 25, 35, 45; 36; 48.

Problem 5.6 A positive 8-digit integer has only 2 different digits. What is the smallest such number that is a multiple of both 5 and 6?

Answer

10000110

Solution

Since $6 = 2 \times 3$, we want a number that is a multiple of $2, 3, 5$. For divisibility by 2 and 5, the last digit must be 0. One of our digits is 0, and if we want the smallest number, we should try 1 as our other digit. To be divisible by 3, the sum of the digits must be divisible by 3, so we need at least 3 ones. The smallest number (with three 1's and five 0's) is thus 10000110.

Problem 5.7 There are four pupils whose ages are four consecutive integers. The product of their ages is 5040. What are their ages respectively?

Copyright © ARETEEM INSTITUTE. All rights reserved.

5 Solutions to Chapter 5 Examples

Answer

7, 8, 9, 10

Solution

Find the prime factorization of $5040 = 2^4 \times 3^2 \times 5 \times 7$. We then form four consecutive natural numbers from these prime factors. The only way to do so is $7 \times 8 \times 9 \times 10$, so the ages are $7, 8, 9, 10$.

Problem 5.8 Barry wrote 6 different numbers, one on each side of 3 cards, and laid the cards on a table, as shown. The sums of the two numbers on each of the three cards are equal. The three numbers on the hidden sides are prime numbers. What is the average of the hidden prime numbers?

Answer

14

Solution

Since the sums of the two numbers on each card are equal, the hidden numbers of these three cards must contain at least one even and one odd numbers. However, since the hidden numbers are all primes, one of them (the even one) has to be 2. The only way it is possible is that the hidden number on the back of 59 is 2. Then the sum is 61, so the other two hidden numbers are $61 - 44 = 17$ and $61 - 38 = 23$. The average: $(2 + 17 + 23)/3 = 14$.

Problem 5.9 A five-digit number consists of distinct digits. It is divisible by 3, 5, 7, and 11. Find the greatest five-digit number that satisfies these conditions.

Answer

98175

Copyright © ARETEEM INSTITUTE. All rights reserved.

Solution

The five-digit number should be divisible by $3 \times 5 \times 7 \times 11 = 1155$. The largest 5-digit multiple of 1155 is $1155 \times 86 = 99330$, but it has repeated digits. So we try $1155 \times 85 = 98175$, which consists of distinct digits.

Problem 5.10 In the multiplication problem below, A, B, C and D are distinct digits. What is $A + B$?

$$\begin{array}{r} A\ B\ A \\ \times C\ D \\ \hline C\ D\ C\ D \end{array}$$

Answer

1

Solution

It doesn't matter what C and D are, and the multiplication shows that $\overline{CD}$ times $\overline{ABA}$ equals $\overline{CDCD}$, so $\overline{ABA}$ is just 101. Thus $A + B = 1$.

6 Solutions to Chapter 6 Examples

Problem 6.1 Find the greatest common divisor and the least common multiple of

(a) 123 and 1681.

Answer

GCD: 41, LCM: 5043.

Solution

We have $123 = 3 \cdot 41$ and $1681 = 41^2$. Therefore, $\gcd(123, 1681) = 3^0 \cdot 41^1 = 41$ and $\text{lcm}(123, 1681) = 3^1 \cdot 41^2 = 5043$.

(b) $8!$ and 4^3 (Recall $8!$ represents "8 factorial", which means $1 \times 2 \times 3 \times \cdots 8$.)

Answer

GCD: 64, LCM: $8! = 40320$.

Solution

We have $8! = 2^7 \cdot 3^3 \cdot 5 \cdot 7$ and $4^3 = 2^6$. Therefore, $\gcd(8!, 4^3) = 2^6 \cdot 3^0 \cdot 5^0 \cdot 7^0 = 64$ and $\text{lcm}(8!, 4^3) = 2^7 \cdot 3^1 \cdot 5^1 \cdot 7^1 = 8! = 40320$.

(c) $3! + 5!$ and $5! + 6!$.

Answer

GCD: 42, LCM: 2520

Solution

We have $3! + 5! = 3!(1 + 5 \cdot 4) = 6 \cdot 21 = 2 \cdot 3^2 \cdot 7$ and $5! + 6! = 5!(1+6) = 120 \cdot 7 = 2^3 \cdot 3 \cdot 5 \cdot 7$. Therefore, $\gcd(3!+5!, 5!+6!) = 2^1 \cdot 3^1 \cdot 5^0 \cdot 7^1 = 42$ and $\text{lcm}(3!+5!, 5!+6!) = 2^3 \cdot 3^2 \cdot 5^1 \cdot 7^1 = 2520$.

Problem 6.2 What is the least common multiple of the first 10 positive integers?

Answer

2520

Solution

Consider the prime factorization of $1, 2, 3, \ldots, 10$. There are four primes: $2, 3, 5, 7$. The highest powers of each prime is 2^3, 3^2, 5, and 7. So we simply multiply these powers to get the LCM of the first 10 numbers: $2^3 \cdot 3^2 \cdot 5 \cdot 7 = 2520$.

Problem 6.3 Among the fractions $\dfrac{1}{24}, \dfrac{2}{24}, \dfrac{3}{24}, \ldots, \dfrac{23}{24}$, how many are irreducible? (Irreducible means the numerator and denominator are relatively prime)

Answer

8

Solution

Since $24 = 2^3 \times 3$, the fraction is irreducible if the numerator is not divisible by 2 or 3. There are 8 of those: $\dfrac{1}{24}, \dfrac{5}{24}, \dfrac{7}{24}, \dfrac{11}{24}, \dfrac{13}{24}, \dfrac{17}{24}, \dfrac{19}{24}, \dfrac{23}{24}$.

Problem 6.4 The least common multiple of two numbers is 180 and their greatest common divisor is 30. One of the two numbers is 90. what is the other number?

Solution

$90 = 2 \times 3^2 \times 5$. Also, $180 = 2^2 \times 3^2 \times 5$ and $30 = 2 \times 3 \times 5$. If we denote the other number as $2^x \times 3^y \times 5^z$, we get (by comparing the exponents of each prime number)

$$x = 2, \quad y = 1, \quad z = 1.$$

So the other number is $2^2 \times 3 \times 5 = 60$.

Problem 6.5 True or False. You do not need to formally prove your answers, but given an example with explanation for each.

(a) If m, n are relatively prime and n, l are relatively prime then m, l are relatively prime.

Copyright © ARETEEM INSTITUTE. All rights reserved.

6 Solutions to Chapter 6 Examples

Answer

False

Solution

Let $m, n, l = 2, 3, 4$. Then $\gcd(m,n) = \gcd(2,3) = 1$ and $\gcd(n,l) = \gcd(3,4) = 1$. However, $\gcd(m,l) = \gcd(2,4) = 2$, so m, l are not relatively prime.

(b) Any common divisors of m and n are divisors of $\gcd(m,n)$.

Answer

True

Solution

The common divisors of $12, 18$ are $1, 2, 3, 6$, and all are divisors of $\gcd(12, 18) = 6$. The idea is that $6 = 2 \cdot 3$ (since $12 = 2^2 \cdot 3, 18 = 2 \cdot 3^2$) and all the other divisors use ≤ 1 powers of $2, 3$ in their prime factorizations.

(c) Any common multiples of m and n are multiples of $\mathrm{lcm}(m,n)$.

Answer

True

Solution

$36, 72, 108, \ldots$ are all common multiples of 12 and 18, and they are all multiples of $\mathrm{lcm}(12, 18) = 36$. The idea is that $36 = 2^2 \cdot 3^2$ (since $12 = 2^2 \cdot 3, 18 = 2 \cdot 3^2$) and all the other multiples use ≥ 2 powers of $2, 3$ in their prime factorizations.

Problem 6.6 A certain fraction, $\dfrac{m}{n}$, if reduced to lowest terms, equals $\dfrac{5}{11}$. Also given that $m + n = 80$. What are m and n?

Answer

$m = 25, n = 55$

Solution 1

An unreduced fraction will satisfy $m = 5k, n = 11k$ (for an integer k). So $5k + 11k = 80$, thus $k = 5$. So $m = 25, n = 55$.

Solution 2

Alternatively, we can use trial and error:

$$\frac{5}{11} = \frac{10}{22} = \frac{15}{33} = \frac{20}{44} = \frac{25}{55} = \frac{30}{66} = \cdots$$

and we see that $m = 25, n = 55$ works. Note here that the sum of the numerator and denominator increases by $5 + 11 = 16$ each "step".

Problem 6.7 How many integers between 1000 and 2000 have all three of the numbers 12, 18 and 30 as factors? What are they?

Answer

6: They are $1080, 1260, 1440, 1620, 1800, 1980$.

Solution

We are looking for common multiples of 12, 18 and 30. The common multiples should be the multiples of the LCM. First, $\text{lcm}(12, 18, 30) = 2^2 \times 3^2 \times 5 = 180$. Between 1000 and 2000, these 6 values $1080, 1260, 1440, 1620, 1800, 1980$ satisfy the requirements.

Problem 6.8 How many positive two-digit integers are factors of both 360 and 540? What are they?

Answer

10. They are: $10, 12, 15, 18, 20, 30, 36, 45, 60, 90$.

Solution

$\gcd(360, 540) = 180$. So this question becomes: how many two-digit factors does 180 have?
We can list the answers: 10, 12, 15, 18, 20, 30, 36, 45, 60, 90, totally 10 of them.

Problem 6.9 Three units commonly used to measure angles are degrees (360 degrees

6 Solutions to Chapter 6 Examples

in a circle), grads (400 grads in a circle) and mils (6400 mils in a circle). A right angle has an integer value for all three units which is 90 degrees, 100 grads and 1600 mils. Find the number of degrees of the smallest positive angle which is an integer for all three units.

Answer

$9°$.

Solution

If y degrees is an integer when converted to grads and mils then

$$\frac{400y}{360} = \frac{10y}{9} \text{ and } \frac{6400y}{360} = \frac{160y}{9}$$

must be integers. Hence y must be a multiple of 9. The smallest positive value is $9°$.

Problem 6.10 Adding the same positive integer N to both the numerator and denominator of the fraction $\frac{1997}{2000}$, we obtain a new fraction that is equal to $\frac{2000}{2001}$. What is this number N?

Answer

$N = 4003$.

Solution

The difference between the numerator and denominator of $\frac{1997}{2000}$ is 3, so it is still 3 after adding the same number N. Thus the new number is $\frac{2000}{2001} = \frac{6000}{6003}$. So the value $N = 6003 - 2000 = 4003$.

7 Solutions to Chapter 7 Examples

Problem 7.1 For each of the following sets of numbers, find the GCD and LCM.

(a) 15 and 21.

Answer

3, 105

Solution

$15 = 3 \cdot 5, 21 = 3 \cdot 7$, so $\gcd(15, 21) = 3$ and $\text{lcm}(15, 21) = 3 \cdot 5 \cdot 7 = 105$.

(b) 50, 90.

Answer

10, 450

Solution

$50 = 2 \cdot 5^2, 90 = 2 \cdot 3^2 \cdot 5$. Thus, $\gcd(50, 90) = 2 \cdot 5 = 10$ and $\text{lcm}(50, 90) = 2 \cdot 3^2 \cdot 5^2 = 450$.

(c) 49, 154.

Answer

7, 1078

Solution

$49 = 7^2, 154 = 2 \cdot 7 \cdot 11$. Therefore, $\gcd(49, 154) = 7$ and $\text{lcm}(49, 154) = 2 \cdot 7^2 \cdot 11 = 1078$.

(d) 8, 20, 24.

Answer

4, 120

Copyright © ARETEEM INSTITUTE. All rights reserved.

7 Solutions to Chapter 7 Examples

Solution

$8 = 2^3, 20 = 2^2 \cdot 5, 24 = 2^3 \cdot 3$, so $\gcd(8, 20, 24) = 2^2 = 4$ and $\text{lcm}(8, 20, 24) = 2^3 \cdot 3 \cdot 5 = 120$.

Problem 7.2 Product of GCD and LCM

(a) What do you notice about the quantity $\gcd(m, n) \times \text{lcm}(m, n)$?

Solution

$\gcd(m, n) \times \text{lcm}(m, n) = m \times n$.

(b) Explain why the observation in part (a) is always true.

Solution

In general, we look at the exponents of each prime factor for m and n. The smaller exponent is chosen for GCD, and the larger exponent is chosen for LCM. If we multiply the GCD and LCM, both exponents occur, and that is the same as multiplying m and n. So we have $\gcd(m, n) \cdot \text{lcm}(m, n) = mn$.

Problem 7.3 How many positive two-digit integers are factors of both 2240 and 2880? What are they?

Solution

$2240 = 2^6 \times 5 \times 7$, and $2880 = 2^6 \times 3^2 \times 5$, thus $\gcd(2240, 2880) = 2^6 \times 5 = 320$. Hence, we are looking for 2-digit factors of 320. We may count all the factors of 320 (answer: 14), and take away the single-digit ones and three-digit ones. We can also list the two-digit factors (the question asks for them anyways): 10, 16, 20, 32, 40, 64, 80. Totally 7.

Problem 7.4 At A.R.Teem Institute, there are 780 students in total, some of whom participate in Math Challenge classes. Among the Math Challenge class students, exactly $\frac{8}{17}$ are 6th graders, and exactly $\frac{9}{23}$ are 7th graders. How many students at A.R.Teem Institute do not attend Math Challenge classes?

Solution

The number of students in the Math Challenge classes must be a common multiple of both 17 and 23. Since $17 \times 23 = 391$, and $391 \times 2 = 782 > 780$, there are exactly 391 students in the Math Challenge classes (including 6th, 7th, and other grades). So the remaining 389 students do not attend the Math Challenge classes.

Problem 7.5 Consider numbers that leave a remainder of 2 when divided by 3, 4, 5, and 6.

(a) Find the smallest such number.

Answer

62

Solution

If the number is N, then $N-2$ must be a common multiple of 3, 4, 5, and 6. $\text{lcm}(3,4,5,6) = 60$, so the smallest such number is 62.

(b) Find the largest such three-digit number.

Answer

962

Solution

If the number is N, then $N-2$ must be a common multiple of 3, 4, 5, and 6. $\text{lcm}(3,4,5,6) = 60$, and in 3-digits, the largest multiple of 60 is 960. Therefore the number we want is 962.

Problem 7.6 In a Math Challenge class, the number of students is between 20 and 30. These students sit around a circular table, and start counting off numbers, clockwise, beginning with 1, and continue until 200 is counted. If the numbers 2 and 200 are counted by the same student, how many students are in the class?

Answer

22

Copyright © ARETEEM INSTITUTE. All rights reserved.

7 Solutions to Chapter 7 Examples

Solution

The number of students is a factor of $200 - 2 = 198$. Among the factors, only 22 is between 20 and 30. So there are 22 students in the class.

Problem 7.7 In a math class, the teacher brings some pencils into the classroom. If he distributed the pencils evenly among the girls, each girl would get 15 pencils. If he distributed the pencils evenly among the boys, each boy would get 10 pencils. In fact, the teacher distributes the pencils evenly among all the students. How many pencils does each student receive?

Answer

6

Solution

The number of pencils is a multiple of 15, and also a multiple of 10, so it is a multiple of $\text{lcm}(15, 10) = 30$. Thus there are $30k$ pencils, where k is a positive integer. For simplicity we may just assume $k = 1$. Any other value of k also works, and does not affect the result (we may try a different value to see the results). So there are 30 pencils. Based on the description, there are 2 girls and 3 boys. So if the teacher distributes the 30 pencils evenly among all 5 students, it will be 6 pencils for each student.

Problem 7.8 Suppose A, B, C are integers ≥ 2 with (i) $\gcd(A, B) = 12$, (ii) $\text{lcm}(A, B) = 396$, and (iii) $\gcd(B, C) = 33$.

(a) What is $A \times B$?

Answer

4752

Solution

$A \times B = \gcd(A, B) \times \text{lcm}(A, B) = 12 \times 396 = 4752$.

(b) Calculate $\gcd(5A, 5B)$ and $\text{lcm}(5A, 5B)$.

Answer

$60, 1980$

Solution

Since we are adding an extra 5 to the prime factorization of both A, B we have $\gcd(5A, 5B) = 5 \times \gcd(A, B)$ and $\text{lcm}(5A, 5B) = 5 \times \text{lcm}(A, B)$.

(c) Calculate $\gcd(11A, B)$.

Answer

132

Solution

Note that $396 = 2^2 \cdot 3^2 \cdot 11$, so one of A or B is divisible by 11. Since $\gcd(B, C) = 33 = 3 \cdots 11$, it must be the case that B is divisible by 11. Therefore, $\gcd(11A, B) = 11 \times \gcd(A, B) = 132$.

(d) What are the possibilities for $\gcd(A, C)$?

Answer

3 or 9

Solution

As in part (c), we know that B is divisible by 11. This implies that A is not divisible by 11. Further, since $\gcd(A, B) = 12 = 2^2 \cdot 3$ and $\gcd(B, C) = 3 \cdot 11$, we know that C is not divisible by 2. Hence, $\gcd(A, C)$ is a power of 3, so either 3 or 9 (both of which are possible).

Problem 7.9 Suppose you live in a society where you only have $4 and $7 dollar bills.

(a) Show that it is impossible to pay for something (using only $4 and $7 dollar bills) that costs 17 dollars.

Solution

Note that 17 is not a multiple of 4, neither is $17 - 7 = 10$ nor $17 - 2 \cdot 7 = 3$. Therefore, it is impossible to pay for something costing 17 dollars.

Copyright © ARETEEM INSTITUTE. All rights reserved.

7 Solutions to Chapter 7 Examples

(b) Show that it is possible to pay for something (using only $4 and $7 dollar bills) that costs 18, 19, 20, or 21 dollars.

Solution

We have $18 = 7+7+4, 19 = 7+4+4+4, 20 = 4+4+4+4+4, 21 = 7+7+7$.

(c) Expand your argument from part (b) to show that $17 is the largest amount you cannot pay for using only $4 and $7 dollar bills.

Solution

In part (b) we showed we can get $18, 19, 20, 21$. Adding an extra $4 bill to each we can pay for $22, 23, 24, 25$. Adding another, $26, 27, 28, 29$. Continuing in this way we see we can pay for any dollar amount larger than 17.

Problem 7.10 There are 9 divisors for number A and 10 divisors for number B. The least common multiple of A and B is 2800. What are these two numbers?

Answer

$A = 100, B = 112$

Solution

First note $2800 = 2^4 \times 5^2 \times 7$. Since A has 9 divisors, it is a square number of the form p^2q^2 (p and q are primes) or the form p^8, based on the formula to count divisors. From the factorization of 2800, A can only be $2^2 \times 5^2 = 100$.

Based on the definition of LCM, B has to have factor $2^4 \times 7$. Notice that $2^4 \times 7$ has exactly 10 factors, we conclude that B is just $2^4 \times 7$ which equals 112.
Therefore, $A = 100, B = 112$.

8 Solutions to Chapter 8 Examples

Problem 8.1 Units Digit

(a) Find the quotients and remainders when 34567 and 45678 are divided by 10?

Answer

$34567 = 10 \cdot 3456 + 7, 45678 = 10 \cdot 4567 + 8$.

(b) Find the units digit of $34567 + 45678$. How does it compare to the units digit of $7 + 8$.

Answer

5

Solution

The units digit of $34567 + 45678$ is equal to the units digit of $7 + 8$.

(c) Find the units digit of 34567×45678. How does it compare to the units digit of 7×8.

Answer

6

Solution

The units digit of 34567×45678 is equal to the units digit of 7×8.

(d) Explain your answers from part (b) and (c).

Solution

Consider the procedures you know for adding and multiplying. The units digit in the answer only depends on the units digits in the original numbers.

Problem 8.2 Everyday Problems with Remainders

Copyright © ARETEEM INSTITUTE. All rights reserved.

8 Solutions to Chapter 8 Examples

(a) Suppose that the date is Saturday March 26th. What day of the week will March 26th be next year? (Assume next year is not a leap year.)

Answer

Sunday

Solution

Note that every 7 days is the same day of the week. Since there are 365 days in a year and $365 = 7 \cdot 52 + 1$, March 26th next year is a Sunday.

(b) Suppose it is 9 o'clock now. What time will it be 100 hours from now, if we ignore am/pm?

Answer

1

Solution

Note that every 12 hours is the same time (with a switched am/pm we don't care about). Since $100 = 12 \cdot 8 + 4$, it will be 1 o'clock 100 hours from now.

(c) What might be a better way to think of a 1000° angle?

Answer

$280°$ or $-80°$

Solution

Note that a full circle is $360°$. Since $1000 = 360 \cdot 2 + 280$, we can visualize a $1000°$ angle as the same as a $280°$ angle. Alternatively, we could also think of it as a $-80°$ angle (since $1000 = 360 \cdot 3 - 80$.

Problem 8.3 Day of the Week Problems

(a) In a leap year, February is a month that contains Friday the 13th, what day of the week is March 1?

Answer

Monday

Solution

February has 29 days in a leap year. If Feb 13 is a Friday, then Feb 27 (14 days later) is a Friday, and Feb 29 is a Sunday. Thus March 1 is a Monday.

(b) Carlos Montado was born on Saturday, November 9, 2002. On what day of the week will Carlos be 706 days old?

Answer

Friday

Solution

If 706 is divided by 7, the remainder is 6. Carlos was born on a Saturday, so after 706 days, it is the same as 6 days after Saturday, thus it is a Friday.

Problem 8.4 Patterns!

(a) Find the units digit of 2^{2016}.

Answer

6

Solution

The last digits of the powers of 2 follow a pattern: $2, 4, 8, 6, 2, 4, 8, 6, \ldots$. The length of the cycle is 4. The exponent 2016 is a multiple of 4 (full cycle), so the units digit of 2^{2016} is the last number in the cycle: 6.

(b) Find the remainder when 2^{2016} is divided by 7.

Answer

1

8 Solutions to Chapter 8 Examples 135

Solution

The remainders when dividing by 7 of the powers of 2 follow a pattern: $2, 4, 1, 2, 4, 1, \ldots$. The length of the cycle is 3. The exponent 2016 is a multiple of 3 (full cycle), so the units digit of 2^{2016} is the last number in the cycle: 1.

Problem 8.5 When 1999^{2000} is divided by 5, what is the remainder?

Answer

1

Solution

First reduce 1999 to 4, since we are dividing by 5, and we only need to work with the remainder. The pattern of the remainders of powers of 4 divided by 5 is: $4, 1, 4, 1, \ldots$. Since 2000 is even, the answer is 1.

Problem 8.6 What is the units digit of $19^{19} + 99^{99}$?

Answer

8

Solution

The pattern of the units digit is: $9, 1, 9, 1, \ldots$ for both 19^{19} and 99^{99}. The exponents are both odd, thus the units digits are both 9, and the units digit of the sum is 8.

Problem 8.7 Each principal of Lincoln High School serves exactly one 3-year term. What is the maximum number of principals this school could have during an 8-year period?

Answer

4 is possible

Solution

We can have 2 principals serving full terms within these 8 years, and one year for the earliest principal, and one year for the latest principal.

Copyright © ARETEEM INSTITUTE. All rights reserved.

Using letters to represent the principals each year: ABBBCCCD.

Problem 8.8 The product of the two 9-digit numbers 404040404 and 707070707 has thousands digit A and units digit B. What is the sum $A + B$?

Answer

13

Solution

Since we only care about the thousands digits and the units digits of the result, only the last 4 digits of both operands are important. So we simply multiply 0404 and 0707, and of course the leading 0s can be omitted and we multiply 404 and 707. Direct computation gets $404 \times 707 = \ldots 5628$ (remember we only care about the thousands and units digits). Therefore $A = 5$, $B = 8$, so $A + B = 13$.

Problem 8.9 A group of pirates went to hunt for treasure. They found a chest of gold coins. They tried to equally divide the coins, but 6 coins were left over. So they picked one pirate among themselves and threw him overboard. Then they tried to divide the coins again, but 5 coins were left over. If the chest held 83 coins, how many pirates were there originally?

Answer

7

Solution

$83 - 6 = 77$, so the number of pirates at the beginning was a factor of 77.

$83 - 5 = 78$, so after throwing one pirate overboard, the remaining number of pirates is a factor of 78.

Looking at the factors of the numbers 77 and 78, we conclude that there were 7 pirates originally, and 6 pirates now.

Problem 8.10 Let D be an integer greater than 1. When each of the three numbers 108, 201, and 356 is divided by D, the remainders are the same number R. Compute the value of $D - R$.

Copyright © ARETEEM INSTITUTE. All rights reserved.

8 Solutions to Chapter 8 Examples

Answer

16

Solution

If two numbers have the same remainder when divided by D, then the difference between these two numbers is a multiple of D.

Take the differences among the given numbers: $201 - 108 = 93$, and $356 - 201 = 155$. Both 93 and 155 should be multiples of D, so D is a common divisor of 93 and 155. Note that $\gcd(93, 155) = 31$, and the only common divisors of 93 and 155 are 1 and 31. We are given that $D > 1$, so D must be 31.

$108 = 31 \times 3 + 15$, so the remainder R is 15. (You may verify that the remainders of the other two numbers divided by 31 are also 15.)

Finally, $D - R = 31 - 15 = 16$.

9 Solutions to Chapter 9 Examples

Problem 9.1 A box contains gold coins. If the coins are equally divided among six people, four coins are left over. If the coins are equally divided among five people, three coins are left over. If the box holds the smallest number of coins that meets these two conditions, how many coins are left when equally divided among seven people?

Answer

0

Solution

If we add 2 coins to the pile, then the coins become evenly divisible by both 6 and 5. The smallest number satisfying that is 30. Taking away the 2 extra coins, we get the smallest number of coins that meets the given conditions is 28. And this number is divisible by 7.

Problem 9.2 Answer the following.

(a) What is the units digit of $3^{215} + 7^{121}$?

Answer

4

Solution

We shall calculate the patterns of both powers of 3 and 7.
The units digit of powers of 3: pattern is $3, 9, 7, 1, 3, 9, 7, 1, \ldots$. The cycle length is 4.
The number 215 is 3 mod 4, so the units digit of 3^{215} is 7.
The units digit of powers of 7: pattern is $7, 9, 3, 1, 7, 9, 3, 1 \ldots$. The cycle length is also 4.
And 121 is 1 mod 4, so the units digit of 7^{121} is 7.
The sum $7 + 7 = 14$, so the final answer is 4.

(b) Find the remainder when $7^{18} + 9^{18}$ is divided by 8.

Answer

2

Copyright © ARETEEM INSTITUTE. All rights reserved.

9 Solutions to Chapter 9 Examples

Solution

Note that $7 \equiv -1 \pmod 8, 9 \equiv 1 \pmod 8$, so $7^{18} + 9^{18} \equiv (-1)^{18} + 1^{18} \equiv 1 + 1 \equiv 2 \pmod 8$, so the remainder is 2.

Problem 9.3 If $m > 1$ and $60 \equiv 70 \equiv 85 \pmod m$, what is m?

Answer

5

Solution

Note $70 - 60 = 10, 85 - 70 = 15$, so $m \mid \gcd(10, 15) = 5$. Since $m > 1$ (and 5 is prime) m must be 5.

Problem 9.4 Find the last two digits of

(a) Find the last two digits of 99^{2016}.

Answer

01

Solution

The pattern of the last two digits of powers of 99: 99, 01, 99, 01, ... Therefore the last two digits of 99^{2016} are 01.

(b) Find the last two digits of 7^{2015}.

Answer

43

Solution

The pattern of the last two digits of powers of 7: 07, 49, 43, 01, 07, 49, 43, 01, ... Cycle length is 4. And $2015 \equiv 3 \pmod 4$. Thus the last two digits of 7^{2015} are the third in the cycle: 43.

Problem 9.5 Find the remainder when

(a) $35^3 + 53^3$ is divided by 10.

Answer

2

Solution

The units digit of 35^3 is 5. The units digit of 53^3 is 7. The sum is 12, whose units digit is 2.

(b) $7^{18} + 9^{18}$ is divided by 8.

Answer

2

Solution

Note that $7 \equiv -1 \pmod{8}, 9 \equiv 1 \pmod{8}$, so $7^{18} + 9^{18} \equiv (-1)^{18} + 1^{18} \equiv 1 + 1 \equiv 2 \pmod{8}$, so the remainder is 2.

Problem 9.6 Modulo 9: Add the digits!

(a) Consider the number 1234. What is the quotient and remainder when 1234 is divided by 9?

Solution

$1234 = 137 \times 9 + 1$.

(b) Consider the sum of the digits of 1234: $1 + 2 + 3 + 4 = 10$. What 10 (mod 9)?

Solution

1

(c) Consider a scrambled and combined sum of the digits of 1234: $34 + 21 = 55$. What is 55 (mod 9)?

Solution

1.

Copyright © ARETEEM INSTITUTE. All rights reserved.

9 Solutions to Chapter 9 Examples

(d) Summarize the results in this problem.

Solution

When working mod 9, we can rearrange the digits however we want, or even break them into pieces, and the sum will always be the same. This works because $10 \equiv 1 \pmod 9$.

Problem 9.7 Suppose A, B, C, D are 4 consecutive natural numbers.

(a) Find the remainder when $A + B + C + D$ is divided by 4.

Answer

2

Solution

Note in some order, A, B, C, D have remainders $0, 1, 2, 3$ when divided by 4. The sum therefore has a remainder of $0 + 1 + 2 + 3 \equiv 2 \pmod 4$.

(b) Suppose you also know that $A + B + C + D$ is a three-digit number between 400 and 440 and $A + B + C + D$ is divisible by 9. Find A, B, C, D.

Answer

102, 103, 104, 105

Solution

We know by part (a) the sum of the four natural numbers is equivalent to 2 (mod 4). Between 400 and 440, the possibilities are: 402, 406, 410, 414, 418, 422, 426, 430, 434, and 438. The only one that is a multiple of 9 is 414. The average of the four numbers is $414/4 = 103.5$, so the four numbers must be $102, 103, 104, 105$.

Problem 9.8 Concatenate the positive integers $1, 2, 3, \ldots, 2016$ to form a new integer:

$$1234567891011121314\cdots 201420152016.$$

What is the remainder when this new integer is divided by 9?

Copyright © ARETEEM INSTITUTE. All rights reserved.

Answer

0

Solution

It is the same as the remainder when we add up $1+2+3+\cdots+2016$, whose result is $\dfrac{2016 \times 2017}{2} = 1008 \times 2017$, which is a multiple of 9. So the answer is 0.

Problem 9.9 Three numbers, 22, 41, and 60 are divided by a positive integer d, and the three remainders are r_1, r_2, r_3 respectively. Given that $r_1 + r_2 + r_3 = 21$, determine the number d.

Answer

17

Solution

d is a divisor of $22 + 41 + 60 - 21 = 102 = 2 \times 3 \times 17$.

If we want the three remainders r_1, r_2, r_3 to add up to 21, d cannot be small numbers like 2, or 3, or 6. But if d is too big, such as $2 \times 17 = 34$, then the number 22 is already its own remainder, and greater than 21. The only possibility is $d = 17$. Verify: $22 \equiv 5 \pmod{17}$, $41 \equiv 7 \pmod{17}$, and $60 \equiv 9 \pmod{17}$. The three remainders $5 + 7 + 9 = 21$ exactly.

Problem 9.10 Initially, the positive integers $1, 2, 3, \ldots, 2016$ are written on the blackboard. Perform the following operation: At each step, three randomly chosen numbers on the board are erased, and replaced with the last digit of the sum of the three numbers. For example, if the numbers erased are 5, 27, and 2001, then write 3 on the board. Or, the numbers erased are 243, 62, and 334, then write 9 on the board. After many steps, only two numbers are left on the board. One is 48. What is the other number?

Answer

8

Solution

The number 48 is never erased, so the other number is the units digit of the sum of all

---

Copyright © ARETEEM INSTITUTE. All rights reserved.

other numbers. Note that this sum only depends on the units digit of the other numbers. The sum of all these units digits (including 48) is thus (grouping them in sets of 10):

$$201 \times (1+2+3+\cdots+9+0) + 1+2+3+4+5+6 = 201 \times 45 + 21 = 9066.$$

Subtracting the 8 from 48 we have $9058 \equiv 8 \pmod{10}$.

www.ingramcontent.com/pod-product-compliance
Lightning Source LLC
Chambersburg PA
CBHW081133170426
43197CB00017B/2851